SLENDER & TRIM

ORNAMENT IS BUT THE GUILED SHORE TO A MOST DANGER~OUS SEA.

SHAKESPEARE

WEARY, STALE, FLAT AND UNPROFITABLE.

SHAKESPEARE

To Viola

my wife.

INDUSTRIAL DESIGN
RAYMOND LOEWY

LAURENCE KING

Published in 2000 by Laurence King Publishing
an imprint of Calmann & King Ltd
71 Great Russell Street
London WC1B 3BN
Tel: +44 020 7831 6351
Fax: +44 020 7831 8356
e-mail: enquiries@calmann-king.co.uk
www.laurence-king.com

A catalogue record for this book is available from the British Library.

ISBN 1 85669 201 9

Printed in Spain by Fournier Artes Graficas, S. A.

TABLE OF CONTENTS

Morning of first communion, 1905.

As French army second lieutenant, 1919.

First civilian suit, 1919.

MY LIFE IN DESIGN

The following is the edited version of many conversations between Mr. Loewy and Peter Mayer, the publisher, tape recorded, transcribed, and annotated from the text between 1978 and 1979 in Paris, Palm Springs, and New York. It was edited in the spring of 1979 with this book in mind, incorporating notes Mr. Loewy made to individual illustrations.

PM: *This is a significant year in the lifework of Raymond Loewy. It may be that you are operating as you have done through five decades but, in fact, this is the fiftieth year of your activity in industrial design. In that period, it's fair to say you have become indelibly associated with the term or concept, with the enterprise itself. Would you want to modify anything you have ever said about the work in light of the present moment.*

RL: No. About thirty years ago I once said that "Industrial design keeps the customer happy, his client in the black, and the designer busy." I still feel that this is a good maxim to keep in understanding what we are about. It may seem facile, but one can infer from it, I believe, a concern for the society, for those who have the responsibility to initiate and produce, and for the profession of industrial design to be understood and involved.

PM: *May I suggest that the idea of the marketplace is not one therefore that passes any special problems or conflict for you as a designer.*

RL: Certainly no philosophical problem. Industrial design exists within the marketplace and helps define it. But I have always liked a phrase someone else once used in a slightly different context, "the shaping of everyday life." The idea of designing industrially without having the marketplace in mind would be both unethical and/or ineffective.

PM: *I think I know where that phrase you have just used comes from: the* Life Magazine *special issue on the contributing events and personalities that shaped this country.*

Cranking the family car, a Renault

You were, if I remember correctly, the only foreign-born creative force mentioned. When did you come to the United States, and why?

RL: I arrived in the States, in New York, in 1919, and, although my associates and I have worked all over the world, I do think that my work is inextricably bound up with America. My early experiences had a great deal to do with what I later found myself doing, and I also discovered that America was a receptive place for a young man with ideas. As a consequence, I am very grateful to the country and hope I have made my contribution.

PM: *What were some of the first things you noticed when you arrived as a young man?*

RL: To put it briefly, I was amazed at the chasm between the excellent quality of much American production and its gross appearance, clumsiness, bulk, and noise. Could this be the leading nation in the world, the America of my dreams? I could not imagine how such brilliant manufacturers, scientists, and businessmen could put up with it for so long.

Although trained in France as an engineer, the first work I got, somewhat by accident in fact, was as a fashion illustrator. Although this wasn't to be my work in life, I can't say I didn't enjoy it. It was the beginning of my involvement with a quite glamorous world, and all my life the *way* I lived was a significant factor in my work activities as well. Most of my early work in fashion was done at *Harper's Bazaar,* and that association was pleasant, as was a certain young woman there who functioned brilliantly as a fashion editor and as a teacher of English for a young Frenchman. But all through the period that I was enjoying this pleasant but superficial career as a fashion illustrator, I imagined that a time would come when I could combine an aesthetic sensibility with my professional background in engineering.

Through the exciting twenties, I never was able to understand why this ingenious new nation did not have a new and fresh look about it. Those who take for granted today's way of life will perhaps understand the colossal efforts that were made to orient a nation of one hundred and fifty million people (at the time) in the direction of everyday

aesthetics of our modern physical world—the later look of the American way of life which has been everywhere copied. Since there was no concept of industrial design, I think historical research will indicate that what was achieved was the work of a handful of pioneers, many of them associates of mine, most of them sadly no longer alive, and all of whom I will try to recall in this interview.

PM: *When you came to America, was there anything in your background that indicated that what we now call industrial design was to be your direction at some later point?*

RL: Perhaps. As a boy I had liked both drawing and physics, and I always abhorred the role of being a spectator. In 1908, when I was fifteen, I designed, built, and flew a toy model airplane which won the then-famous James Gordon Bennett Cup. My parents and I lived in Paris, and I rented a small empty stable, constructed an atelier, built and patented the *Ayrel,* the toy plane that was advertised and merchandised all over France. It sold so well that I soon had an offer for my company and sold it, so that I could get on with my studies. By sixteen I had discovered that design could be fun and profitable, and this lesson has never been lost on me.

PM: *I'm still not quite clear on how you left the world of fashion illustration to become an industrial designer.*

RL: It was as a consequence of the larger world I was moving in. I printed up a card and sent it to everyone I knew. It said, "Between two products equal in price, function, and quality, the better looking will outsell the other." In addition to sending it out, I made an effort to meet every top executive I could, actually visiting them, hoping that, aside from those who would kick me out, one or two might say, "Show me what you mean by that."

In 1929 one did. Sigmund Gestetner, a British mimeograph manufacturer, showed up one day and asked me to design a new model for him. He came to my apartment with the model he was then manufacturing and selling, which looked and even smelled awful. Working in my small living room, modeling in clay on a tarpaulin, I de-

As a teenage schoolboy in Paris, I saw the Brazilian inventor, Santos Dumont, fly his handmade plane about a hundred yards over a polo field in the Bois de Boulogne. Excited by this, I designed and built a scale model powered by twisted rubber bands, and it flew so well that it won the cup awarded by the American tycoon, James Gordon Bennett. I called it the *Ayrel*, patented and manufactured it, sold it successfully, and learned a great deal in the process. It established in my mind some hints of what design, inventiveness, and enterprise might mean in my later life; success became more than an unattainable abstraction.

World War I.

MONOPLAN AYREL (R. L.)

MODÈLES DÉPOSÉS EN FRANCE ET A L'ÉTRANGER

LAURÉAT

DE LA

LIGUE AÉRIENNE

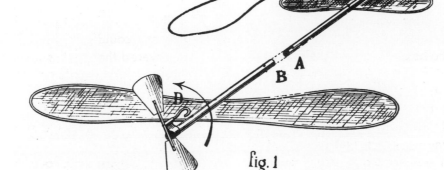

fig. 1

GRAND PRIX

Automobile Club **A**

MONTAGE

1º Introduire le tube A dans le tube B.

2º Tendre le caoutchouc entre les crochets C et D.

RÉGLAGE

L'hélice étant vers le sol, le petit plan est à l'avant, le grand à l'arrière.

1º *Tordre l'aile droite du grand plan, de façon que la face supérieure de cette aile regarde en avant.*

2º *L'aile gauche doit être horizontale.*

Avec un peu d'habitude, on arrive à donner à l'aile droite du grand plan le gauchissement exact qui correspond à la direction que l'on veut imprimer à l'appareil.

LANCEMENT

1º Remonter l'hélice dans le sens de la flèche jusqu'à double torsion du caoutchouc (160 tours environ);

2º Tenir l'appareil de la main droite, l'hélice doit être entre le pouce et l'index et le caoutchouc au-dessus des plans. Le petit plan est donc en avant.

Incliner légèrement l'appareil vers le haut et lui donner une légère impulsion.

Le *Monoplan Ayrel* s'élèvera et fera un vol plané de 150 mètres environ.

Mettre de temps à autre une goutte d'huile sur l'axe de l'hélice.

ANWEISUNG

1. Man stecke das Rohr A in das Rohr B hinein.

2. Dann spanne man die Gummischnur zwischen den Haken C und D.

RÉGULIRING

Die Schraube ist gegen die Person die den Apparat handhabt, gerichtet; die kleine Tragfläche ist vorne, die grössere hinten.

1) Man gebe dem linken Flügel der grossen Tragfläche eine solche Biegung, dass die obere Fläche dieses Flügels nach vorne schaut.

2) Der linke Flügel muss horizontal sein.

Mit ein wenig Uebung gelingt es dem rechten Flügel der grossen Tragfläche die genaue Biegung zu geben, nelche der Richtung die man der Flugmaschine geben will, entspricht.

ABSCHNELLUNG

1) Man ziehe die Schraube nach der Pfeilrichtung auf bis zur doppelten Drehung der Gummischnur (Ca. 160 Umdrehungen).

2) Man fasse die Flugmaschine mit der rechten Hand, so dass die Schraube sich zwischen dem Daumen und dem zeigefinger befindet und die Gummischnur über den Tragflächen. Die kleine Tragfläche ist mithin vorne.

Dann halte man die Maschine etwas nach oben gerichtet und gebe ihr einen leichten Anstoss zum Abflug.

Der Monoplan Ayrel steigt dann in die Höhe und führt einen Gleitflug von Ca. 150 Metern aus.

REMERKUNG

Von Zeit zu Zeit muss man die Schraubenachse ein bischen ölen.

DIRECTIONS

a) Introduce the tube A into the tube B.

b) Affix the rubber to the two hooks C and D.

REGULATION of the APARATUS

The screw pointing towards the manipulator, the smaller plane is in front, the larger one behind.

1) Bend the right wing of the larger plane, until the upper side of this wing points forward.

2) The left wing has to be horizontal.

With a little practice one is able to bend the right wing of the larger plane exactly corresponding with the direction desired.

Apply a little oil occasionnally on the axle of the screw.

Be careful that the rubber is always above the planes.

a) Turn the screw from right to left in the direction of the arrow until double torsion of the rubber (about 160 turns).

b) Holt the aeroplane with your right hand, the screw between your thumb and forefinger, the smaller plane pointing forward.

Give the aeroplane a slight impulse upwards, when you let it go off.

If properly handed. it will then fly about 150 mètres.

PIÈCES DE RECHANGE

Hélice Kalium nº 1	1.00
" " nº 2	1.25
Hélice Cellulo nº 1	1.00
" " nº 2	1.25
Hélice Fibre nº 1	1.00
" " nº 2	1.25
Moteurs nº 1	0.75
" nº 2	1.00
" nº 3	1.50

HIS ROYAL HIGHNESS, THE PRINCE OF WALES
IN THE FULL UNIFORM OF A LIEUTENANT
OF THE ROYAL NAVY, 1913

Edward
Duke of Windsor

The Waldorf-Astoria
A HILTON HOTEL
NEW YORK, N. Y. 10022

April 1st, 1971

Mr. Raymond Loewy
600 Panorama Road
Palm Springs, California

Dear Raymond,

Thank you for your letter of March 21st
and the picture which followed me from Paris.
With much pleasure I return it signed as you
request.

What a long time ago were those days with
the Breteuils. Almost 60 years I am afraid. Still
I hope we can weather a few more!

The Duchess joins me in sending you and Viola
our best wishes and we hope you have spent a
pleasant winter in Palm Springs.

With kind regards,

Sincerely yours,

Edward
D f W

The young Duke of Windsor. His letter thanking me
for sketch on the right.

My office at 500 Fifth Avenue.

Uriane, St. Tropez.

Diving at St. Tropez.

George Carpentier and friends.

The first *Ayrel.*

signed his new machine for him in less than a week. He liked it, took the design back to London, built it, and sold it so well that not only did his company prosper but he essentially kept the same model unchanged for forty years. Through my French company, we are still designing new machines for Gestetner nearly fifty years later! I don't know if I was the first to model in clay to produce an industrial shape, but clay modeling to produce a mock-up was not used by anyone I knew at that time. And because Gestetner needed the design so quickly, there was no way to work in steel. I had to work with my hands, like a sculptor. I kept as close to the skeleton as possible to be efficient. And the end result was a simplified form, designed to prevent the oil, ink, and paper from clogging the works, a much easier machine for a secretary to maintain.

PM: *In other words, what we sometimes talk about as an improvement in the visual appearance of the machine was actually a functional step forward.*

RL: That's right. And that is in the nature of industrial design.

PM: *Let me ask a different question. Since you've just mentioned the clay modeling, is this a system that is still used in industrial design?*

RL: Oh, it's used everywhere. From General Motors to the smallest manufacturer.

PM: *So when work has to be done quickly today, clay modeling is used, just as when you used it for Gestetner in 1929.*

RL: Absolutely. NASA sent me a letter after I worked for them telling me that millions of dollars had been saved by the use of clay, which NASA for some reason seems not to have used before.

PM: *Perhaps this is a good moment to ask you about the Sears Roebuck Coldspot refrigerator of 1934. If I remember correctly, it was the first time an appliance was advertised to the public as much for its beauty as its utility.*

RL: That's true. I had nothing to do with the advertising, but obviously the Sears people, having asked me to design the refrigerator and I suppose liking it, chose to mount a campaign behind it. They noticed that sales quadrupled. This phenomenon might be said to mark the first work commissioned as industrial design in America.

PM: *We've discussed how design at its best appeals aesthetically and adds a functional component. Would you say that a portion of your work, part of the Loewy look based on aerodynamic principles, particularly in the case of all kinds of vehicles, has been a kind of sheathing of the innards?*

RL: Well, to me, sheathing is not really the expression. A well-designed machine requires certain forms, three-dimensional forms. Machines, to begin with, do not optimally make it. It would be a—can you imagine a typewriter without a shell, the noise and the dirt? Or a locomotive without a form? So, I don't call that sheathing. What you call sheathing is a natural form that the machine takes by itself.

PM: *It is the shape, you mean. It's not superimposed.*

RL: In other words, whether the sheath is on it or not, the machine underneath it already has that shape. But a locomotive, for instance, would be terribly difficult to maintain if the machine were exposed. And it would look awful, too. Because, you know, locomotives—I've seen better-looking things. At high speeds, it would be far less efficient. The smoke would be all over the place. So we design locomotives so that the smoke rises above the cab. Think of airplane wings. They create suction over the front end. What you call sheathing is really the self-expression of the machine; when design and engineering work together toward a concern for shape (whether because of a physical principle or beauty or even a safety consideration), there is as much working backward *from* an optimal form *to* mechanics as there is *from* the machinery *to* what you have called sheathing.

PM: *I think you ought to pursue this line, Mr. Loewy, be-*

In 1934, the Metropolitan Museum of Art in New York made this mock-up of my office.

cause I don't believe there's a great understanding on the part of the public with respect to those design elements that are pure appearance and those that have a functional role to play.

RL: I once said that the most difficult things to design are the simplest. For instance, to improve the form of a scalpel or a needle is extremely difficult, if not impossible. To improve the appearance of a threshing machine is easy. There are so many components on which one can work. Another thing: Naturally, sheathing is dictated by the form of the working parts of the machine. It follows its volume and shape closely. But there are cases . . . new machines are being developed in which the components . . . one of the items perhaps . . . can be rearranged to produce an end result more reasonably harmonious, sometimes even more efficient, more compact, or perhaps less costly, easier to maintain, easier of access for maintenance and repair. Or even to make packing and shipment easier and less costly. In such cases, the form may dictate the concept of the engineering part. That is the application of the mathematical principle of combinations, permutations, and arrangements. My engineering training has been helpful I believe, and it often gave our organization an edge over competitors.

PM: *You once said to me that of many creative forces important to you, the names Shakespeare, Seurat, Monet, Conan Doyle, Picasso, Nureyev, Chanel, Archipenko, Maugham, Saki, Dali, Cortázar, Diaghilev, Escoffier came first to mind. Quite an eclectic range. You said that you felt they produced great work and beauty because their cardinal element (from a structural point of view) was surprise based on logic. How do you feel this relates to industrial design?*

RL: Oh, I could have added the work of hundreds of others who have been major influences on me. I've never tried to be consistent. I think possibly there is a definite thread through my work. I've read the phrase "it has Loewy flair" about my work (as well as work that imitates mine) without ever knowing precisely what was meant. In other words, what is consistent in my work has been consistent without my intending it so. As I've said, I've also

sought to surprise. Whatever shackles I've had, I've put on myself. I've always resented restrictions. I believe most in an educated intuition, what you get through profound experience. But you must have background in your work, and you must be involved; ideas emerge from background.

I remember when the chief engineer of Porsche in Stuttgart asked me, "Loewy, how did you wind-test the Avanti?" I said, "Why do you ask?" "Well, we know a little here about streamlining and your Avanti is almost perfect, no parasitic noise at high speed, skin friction reduced to practically nothing." I said, "I didn't test it at all." He couldn't believe it. "No," I said, "I did it by feel and design intuition."

PM: *Without applying mathematical principles?*

RL: Well, I knew that the right form was necessary. And a knowledge of the rules.

PM: *This seems to be a kind of precept from the point of view of young people and the world of industrial design.*

RL: Well, on that score, I have to say I believe in natural talent; if you don't have it, do something else. Don't bother about sociology and ideology. What you need is a good knowledge of engineering, paper, pencils, and a slide rule, common sense, and a respect for the arts of the past. And an ability to cooperate.

PM: *Is there a strategy as well?*

RL: Only in a general sense. But a general sense is useful. I believe one should design for the advantage of the largest mass of people, first and always. That takes care of ideologies and sociologies. I think one also should try to elevate the aesthetic level of society. And to watch quality control always, while insisting others do, too. Design simplicity is the essence; the main goal is not to complicate further the already difficult life of the consumer. Our goal is to give him some peace of mind. Junky stuff is consumer murder. If that is a general strategy, perhaps one could say that there are some specific tactics. For example, designers ought to get involved at the inception of a program, not in the middle or the end. Good design is not an applied

Scenes from La Cense.

La Cense "bistro."

Tierra Caliente in Palm Springs.

veneer; it must be integral to the project. And the following things should be treated respectfully: function, ease of operation, maintenance, cost of upkeep, storage, cost of manufacturing, packing, shipping, display, safety, fail-safe operations, and clear, simple, operational-instruction leaflets—all these and more are involved in doing the job properly and thinking of the consumer first.

PM: *You've sometimes spoken of parasitical factors.*

RL: Yes, there are four that are always with us, which we must seek to eliminate: noise, vibration, air or water resistance, villainous smells. There is also the human level.

PM: *What do you mean?*

RL: I mean at the level of the cooperative nature of the industrial design activity. The designer must think always of the unfortunate production engineer who will have to manufacture what you have designed; try to understand his problems. Also, give those who help you proper recognition, mention their names to top management.

PM: *You are in such good health and you are so active that I wonder how you have been able to maintain your work level for such a long time.*

RL: Well, this is as good a chance as any I may get to speak personally. As I mentioned earlier, the way I live is part of my work. Perhaps it really is for everyone, but in my case it is absolutely central. You won't mind if I reminisce?

I married my bright and lovely Norwegian-American wife, Viola Erickson, thirty years ago and she's about thirty years younger than I am. As you know, I've always loved speed. For instance, when we lived at Sands House, my Long Island home which Captain Sands built in colonial days, normally only an hour's drive from New York, either I drove too fast or my Stutz was a red flag to the tough Long Island motorcycle cops. Business was thriving, and I designed a fast cruiser for myself, had it built in California and shipped to Long Island: the *Media Luz,* a forty-foot, forty-knot, twin-screw affair. With no cops on my tail, I went back and forth to New York in about fifteen minutes, usually with a drink close at hand.

I often had guests not from the business world on the boat for excursions out on Long Island Sound. I remember one weekend with the French heavyweight boxing champion Georges Carpentier aboard, only a day before his fight with Jack Dempsey.

Sands House weekends brought together attractive men and women from the magazine, fashion, and film worlds. We had an international scene, with people arriving regularly from Paris, Rome, London, or Beverly Hills. Those were elegant days, and I still see us playing badminton at night on a floodlit court. The chef was excellent, the bar well stocked, (mint juleps the specialty), plenty of popular music in the background. The weekends raced by. There was a large blue-tiled swimming pool at the edge of the lawn, which itself led down to a sandy beach. On one side of the pool was a small underwater lounge with a thick plate-glass window, through which you could see many of the first bikinis ever worn!

My love affair with motor vehicles of any sort was already strong; we kept a powerful Harley-Davidson motorcycle and two custom Lincoln Continental town cars.

The most important thing about Sands House as a memory is that Viola—before she was my wife—often came out with friends over the weekend. At the time, she was the PR Director at Philip Morris.

I'm sure my life style was an easy target for other designers. It probably made some of them resentful and critical, because luxurious living didn't seem to interfere with the firm's increased reputation and large output. But a good life has been as important to me as my work; in fact, the two of them are bound up in each other for me, and I hope that my work in industrial design, that industrial design itself, has made life better for others as it has for me.

I can also say that while Viola and I both like some continuity and stability in our private life, we always resented routine and craved diversity. We are still ready to travel extensively at a moment's notice anywhere for either work or pleasure. We like constant change of pace, new rhythms, new places and climates, and even new friends. We're interested in other people in other civilizations, other sounds and other scents and flavors—all this which is part of our life is actually our raison d'être. We want to observe, be receptive, friendly to new attitudes, and we

want to remember. Every kind of offbeat attitude and expression, whether of dress, wit, or humor, fascinates us. Many of our colleagues in industrial design prefer a more stable life style, but it's not for us. Viola, I should say, is deeply involved in the administration of our European operation and has been for years.

PM: *Is that where you mostly live now?*

RL: We now spend our time in several places, but mostly Palm Springs and Paris. We also have a country home outside Paris. We continue to live in both a social and an industrial world. But I was never a stranger to factories, or to the people who worked in them.

PM: *Your design experience on an international level seems to have also provided a social setting. Your connection, for example, with Malraux, the Windsors . . .*

RL: Yes, my friendship with the Windsors was strong from the time of our first meeting. Viola and I were at their home in Paris a few years ago, and after dinner, as we walked over to the library, Edward put his arm around my shoulder and said, "You know, Raymond, our lives have been event-

The duplex-penthouse office at 580 Fifth Avenue, 1938.

Tierra Caliente in Mexico.

October 31, 1949.

la Erickson Loewy following our wedding, 1948.

TIME

THE WEEKLY NEWSMAGAZINE

DESIGNER RAYMOND LOEWY
He streamlines the sales curve.

he Loewys at home.

ful, extraordinary in so many ways." I agreed with the former king of England, although I wasn't quite sure what he was getting at. By the way, he was a lovely, highly sensitive man, often misunderstood. Our first meeting was curious: while flying from London to Paris in the late twenties on a biplane transport, the then Prince of Wales had the seat in front of me. Planes were, of course, much slower then; a strong headwind would nearly stall the aircraft, and the lengthy flight was an occasion to chat with the nearby passenger. I made a rough sketch of him then, which he seemed to like, and it marked the beginning of a long friendship.

As for Malraux, I don't remember our first meeting, but we knew him over the years. In the period that he was France's minister of cultural affairs, we often discussed a plan of applying industrial design to France, similar to the later Kennedy plan. When Malraux came to New York in May 1962, as the guest of honor at a memorable banquet, Ambassador Alphand, the president of the French Institute, and Viola and I received him. We pursued our earlier discussions but, like Kennedy, Malraux died before the projects could be undertaken. He greatly admired America and took every occasion to make it better understood by the French.

PM: *And Kennedy? How did you come to know him, and was there an industrial design connection?*

RL: Well, yes. I was unimpressed by the gaudy red exterior markings and what seemed to me the amateurish graphics of *Air Force One,* which I saw in 1963 when President Kennedy visited Palm Springs. A friend of mine, General Godfrey McHugh, air attaché to the president, asked me, "Why don't you suggest a new design idea?" He said that a new *Air Force One* was being built and suggested that I redesign the markings. In an unofficial way and without compensation, I agreed to try. After JFK and Godfrey returned to Washington, I received a phone call from my friend telling me that the president wished to meet with me. I flew to the White House, the beginning of a remarkable relationship. *Air Force One* was Kennedy's baby. He and I discussed what it should look like, and he asked me to come back soon with some sketches. A week later I showed him four different versions, large color drawings about thirty inches wide, of the exterior markings. In every case I had replaced red by a luminous ultramarine blue. There were also various versions of simple classic typography. Kennedy became increasingly interested and suggested slight modifications. For our appointment, I brought along sheets of colored paper, scissors, razor blades, and rubber cement. Since his desk in the Oval Office was relatively small, we just sat on the floor cutting out colored paper shapes and working out various ideas. We had three such sessions lasting about an hour each, and he finally approved a design quite similar to one of my early suggestions.

Our offices at 485 Madison Avenue.

Jacqueline Kennedy, remembering my friendship with her husband, asked me to design the memorial postage stamp issued on May 29, 1964, the date of his birthday. I developed several designs based upon a snapshot she had selected, and the final stamp was established with her collaboration. Since our design was naturally not on a fee basis, as a mark of appreciation I received dedicated first-day-of-issue canceled letters from various members of the Kennedy family.

Staff:
This is Jackie's choice.
Let's try darker shades.
~ me first proofs. *[initials]*

Try darker blue.

1

Darker blue grey
(Slate.)

2

Try This.

3

[signature]
march 1964

UNITED STATES OF AMERICA

Japanese visit including Loewy critique of Japanese manufacturers at trade seminar.

JFK enjoyed his plunge into industrial design so much that he told his secretary, Mrs. Lincoln, that we were not to be disturbed while we were "working." On one occasion, when I collected all my stuff and left him, I found the secretary of state patiently waiting outside; he smiled at me and we shook hands, he with a twinkle in his eyes. Incidentally, JFK asked me to place a large presidential seal on the fuselage. General McHugh pointed out to him that this would not be possible; the seal should not appear when the president was not aboard, which happens occasionally. JFK then called the secretary of the air force, whose answer was similar to McHugh's opinion. The president said that he would get back to me on this later. After a few days, McHugh called and gave me a cheerful "go ahead." He told me later that JFK had quite a tough time with the governmental bureaucracy getting the seal on.

Then JFK asked me if I would decorate his and Jackie's stateroom aboard the aircraft, which I did. He said—on a Friday—that he wanted, between the twin beds, a pale blue rug with an American eagle in the center of an oval formed by thirteen stars. I flew back to New York and made a sketch. There a friend of mine, Eddie Fields, had two rugs woven by hand *over the weekend* (by his special process), and I brought them back to the president on Monday before lunch. JFK could hardly believe it; he liked them and gave me one for my home. It is now in front of the fireplace at La Cense.

JFK flew to Russia aboard the new *Air Force One*. Years later the Russians asked me to design the interior of their supersonic aircraft, the Tupolev; but I couldn't just then since we had been retained at that time by Air France to work on the Concorde interiors. When I next met JFK, he

suggested developing other projects together. He asked me for ideas. My view was that industrial design principles could be applied nationally; the physical appearance of the country could be aesthetically upgraded, but only with governmental leadership. The task represented an enormous project, aborted by JFK's assassination.

PM: *Your design connections extended as well, didn't they, to other members of the Kennedy family?*

RL: Yes. In 1967, Ethel Kennedy and Pat Lawford wished to raise funds for retarded children. Our design group offered to design a candle and holder and the campaign's collection box, as our donation. The candle and holder sold in large numbers. We had connections with Johnson, as well. In 1967, under the Johnson administration, I was asked to become a member of the President's Committee for the Handicapped. After a comprehensive study of the items produced by the disabled themselves, I found many quite unnecessarily unattractive and unsalable. This being their main source of income, improvements were essential. I enrolled the benevolent assistance of the entire American industrial design profession. The response of colleagues was a credit to their concern and sense of responsibility. Designers all over America supplied handicapped people at no cost with good designs carefully adapted to various forms of physical disability. The results were outstanding; both sales and selling prices went up considerably. Viola (who was also involved) and I were invited to the White House, where Mrs. Johnson, the president of the Association for the Handicapped, and the wife of the secretary

Representing American delegation, Armistice Day ceremony, Mont Valérien, Paris.   Viola made honorary sergeant of Loewy regiment.   With Sir William Rootes.

of labor noted our and our colleagues' organizational contribution.

PM: *What was the* Star of Hope, *I seem to remember that?*

RL: In 1956 I proposed that a small capsule, to be called the *Star of Hope,* be inserted in earth orbit. Bearing the flags of all nations in our world, it would be constantly broadcasting the message:

PEACE ON EARTH
GOOD WILL TO MEN

Its flashing blue light would have made it visible in the sky at night.

Lyndon Johnson, then senator, liked the idea and read my plan into the *Congressional Record* and proposed its adoption to Congress.

PM: *These things are far afield from the industrial design image of Raymond Loewy. But perhaps what your life suggests is the diversity of interest and achievement. Are you also suggesting that by constant diversity, constant cultural exchange with people from every walk of life, including always the people who have to execute what you have designed, the industrial designer does not become insular but gets . . .*

RL: Yes, an overall picture of the marketplace in all its aspects.

PM: *"The marketplace" again.*

RL: I know it's a rather corny expression, but I've tried to find another one; "marketplace" will have to do; it seems best to define involvement in a real world.

PM: *Perhaps that's why Charles Luckman of Lever Brothers said of you, "Loewy keeps one eye on the imagination and one eye on the cash register."*

RL: If he said that, I think he meant his cash register as well as my own. Business matters, and I don't mind stressing it. Our Loewy teams became the largest industrial design organization in the world by satisfying the needs of real people in the real world. When I arrived in this country it took me a little while to get going, but once I did, we never looked for clients or lacked for them.

PM: *How do you mean?*

RL: Our clients came in over the transom—they sought us out. Our studio grew by itself almost, although we worked extremely hard. We obtained many long-term contracts with large corporations and kept to high standards; we were known as reliable, sound designers with a deep sense of the realities. Above all, we understood marketing and merchandising and how to cooperate with our clients' engineers. Once while engaged in a major project I was phoned at Sands House in Sands Point, Long Island, by the secretary to the president of the Pennsylvania Railroad, the employing firm. He wished to see me personally, as soon as possible, on an urgent matter. It was Friday afternoon, and, as he was cruising on the Hudson River, he would not be accessible until Tuesday.

I knew his boat and decided to hire a two-seater seaplane in Port Washington. We flew over the Hudson at low altitude, and we spotted the boat near Albany. We landed alongside and discussed what my client had in mind. When he returned to Philadelphia on Monday, his instructions were already being followed. We were conscientious, and went after the business, gave our all for

26

Mayor Daley recognizing design assistance at Chicago's International Exposition, 1950.

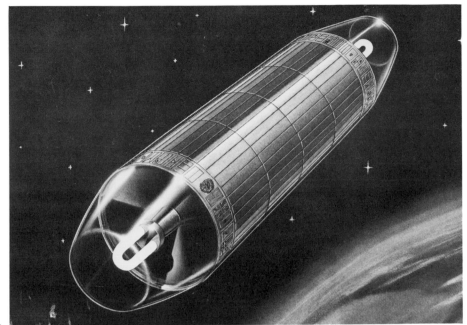

Star of Hope, 1956.

the job, and this became known about us. We have not changed. We always experienced growth, excitement, success with young and talented people who had a sense of humor. We made money and enjoyed spending it. But we also established profit sharing within the firm and did not have to sell ourselves to clients, even when we grew large and had high overheads.

PM: *Tell me, is there an advantage to a large studio over a small studio?*

RL: Yes, there are certain advantages, but conditions have changed. What was true when I talked about the large organization about twenty years ago no longer holds true. Now my organization is centered in Europe. I have three firms; they are small, compact, and I will not let them expand any further. Fifty colleagues—I don't like the word *employee*—fifty collaborators—fifty in each office is maximum. We're up to forty-five or forty-eight in each, and I'm going to keep it there.

PM: *In other words, different sizes are appropriate in different decades, conditions for design vis-à-vis world change.*

RL: That's right.

PM: *Right now, the time calls for more compact organizations than twenty or thirty years ago.*

RL: Besides, and I'm talking about France—if the government moves to the left, there could be a new law passed whereby any organization with more than fifty employees will have to share the management voice or role with the employees. This may be fine in some industries or businesses, but certainly not in industrial design. There must be a head, someone who gives direction . . .

PM: *Vision?*

RL: A vision. Establishes a target, and everybody's got to shoot for that target, and not wander around, right and left.

PM: *Given the fact that industrial design organizations would be of different sizes in different times, nevertheless perhaps there is an ideal way to organize an industrial design studio, regardless of size. The same number of departments always? Regardless of size, is there a way to organize an industrial design studio?*

RL: I think I understand you. In fact we are already going in that direction. For instance, there are excellent free-

Japanese influence on our life at Uriane, St. Tropez

lance designers, so you can keep a skeleton of leaders who instruct the free-lance young folks who do very good work, providing they're watched. Another trend we're establishing now is to push forward with fewer sketches, designs, and renderings . . . to go directly into three dimensions.

PM: *Into models.*

RL: Models, mock-ups, scaled . . .

PM: *Forgive the naïveté of the question. I thought it possible that maybe fifty percent was in meeting people and getting contacts.*

RL: We haven't got a single, not one, salesman.

PM: *I see.*

RL: Because nobody can sell industrial design *but* an industrial designer. We tried salesmen. We tried them in New York. We had all kinds, very high-priced fellows they were. Absolute flop.

PM: *I suppose you mean that the important thing is to keep the emphasis on the creative, and then work flows into the studio.*

RL: That's right. Over the transom. By emphasizing quality design.

PM: *When I ask you to tell me how a design studio should be organized, the answer seems to be: Keep the emphasis on the design part of it, forget sales, and work comes in over the transom. But that's not quite the question I was asking about the studio itself, or the organization itself. Is there a marketing part, a public-relations part?*

RL: Well, sure. We have a director. Director general in France. Managing director in London. He's got an assistant. He has a supporting staff. An accounting department. It's quite important, the accounting department. A CPA as a consultant. Then the actual design team, without which there would be nothing.

PM: *What would be the largest of the different departments?*

RL: The design studio.

PM: *The design studio the largest. The others are service, satellite departments?*

RL: Service, yes, and public relations.

PM: *You would say that any design organization ought to keep itself lean and concentrated, that the design department should remain the largest, and that the bureaucracy should be kept to a minimum around the design center.*

RL: Yes, and if you choose people well—the bureaucracy, and pay them well—a small number can do a very good job. But I would say that, as a guess, eighty percent of our staff is in the design part of the studio.

PM: *What brought about your decision to close Raymond Loewy International in New York?*

RL: I built my company for product, packaging, and architectural design. As I grew older, my talented partner, William Snaith, was becoming exclusively a store designer. I was over seventy-five and less actively engaged in the management; I had completed what for me was my most important assignment: *Skylab.* I can't deny that it concerned me that, as Bill Snaith developed the store division, the product- and graphic-design part of our work inevitably declined. It reached a point where the firm was doing very little outside of store design, and this had never been my particular field. However, my two European organizations were thriving in the original directions that I mentioned above. It was also quite obvious that designers all over Europe were making great strides, becoming leaders in the industrial design profession. So, I sold my company in New York but retained the property of the name, Raymond Loewy International, throughout the world. In this way I could continue to expand on the initiatives closest to my heart.

PM: *Mr. Loewy, I wanted to ask about logotypes. It seems that your logotypes, from International Harvester to Shell, have made a very bold impression on consumers. What do you look for, or what do you seek to achieve with a logotype?*

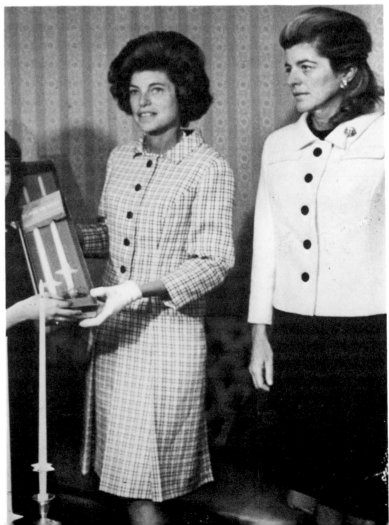

RL: That can be defined very simply, at least in my case. I'm looking for a very high index of visual memory retention. In other words, we want anyone who has seen the logotype, even fleetingly, to never forget it, or at least to forget it "slowly," you might say.

PM: *For example, the double x in Exxon?*

RL: Well, that's an example of new subliminal logotype design. I have an anecdote about that. I had dinner here in Palm Springs a few years ago with Mr. and Mrs. Donald Stralem. A dinner party. Black tie—ridiculous in Palm Springs. My neighbor at my table was a lovely young lady, and she asked me suddenly: "Why did you put two x's on Exxon?" I asked her: "Why ask me?" She said, "Because I couldn't help seeing it." I replied, "Well, that's the answer."

PM: *The range of your design is exceptionally broad. I can't conceive of too many areas having to do with human use that you haven't been involved with. For example, consumer appliances, architecture, transportation—boats, cars, planes, buses, even helicopters. You've also been involved in packaging, and one could go on. Let me ask now—has this variety been accidental or intentional? Absolutely necessary? Has it led to a cross-fertilization of ideas?*

RL: Well, I would say it's been accidental in very few cases; it's ninety percent intentional. Then you ask me about cross-fertilization.

PM: *You say most of it has been intentional, and I was asking whether projects done in transportation have been useful in terms of packaging or architecture or consumer appliances.*

RL: Yes, very often, especially in my case. I've been highly conscious from the start of the whole range of business. We sought work in so many industries so that we could bring design information from one industry—an item or a device, say—to another.

PM: *On a different point: To my knowledge, you are the only designer ever to get a cover story in* Time. *It cer-*tainly helped create a larger public consciousness for design activity. Did you see it as an opportunity when Time was working on it.*

RL: I did see the *Time* story as an opportunity. Their researchers spent several months, I was told, working on it. They asked me about designs I considered successful, designs I considered not up to the mark. I didn't feel comfortable with this line of questioning and suggested they inquire not of me but of present and former clients. Every designer has solutions that are more or less successful than others. But our firm never, I believe, delivered a project to a client that was not an improvement on the situation. I spoke to *Time* instead of projects that I had refused to accept, such as the redesign of burial caskets. I don't like caskets at all. Much more recently a manufacturer who had developed a new type of hand grenade far more murderous than present models thought that I could design it so as to *further* increase fragmentation and, therefore, the number of casualties. He invited me to lunch at the Palm Springs Racquet Club and mentioned it rather slyly between the chef's salad and the lime pie. I was very shocked by the casualness of the proposal, just as if he had asked me to design cosmetic bottles; I left him with his chef's salad, and he never called me again.

PM: *At various times, you've mentioned to me the problem a designer has with respect to the acceptability of progressive design within the expectations of a consumer society.*

RL: Yes, not only the designer but also the manufacturer frequently has problems by introducing a product that leapfrogs either a need or an aesthetic development. Beauty, style, design, utility—there is a relative aspect to all of these words, even if in one individual there seems to be a subjective pure truth to his understanding. Both designers and industrial leaders, while attempting to be progressive, perform an intermediary role, and there are numerous cases of products that were introduced which were rejected in their time only to reappear later after an initial failure. The function of the product satisfied too small a portion of the population to market it successfully, or the general state of the technology was still lagging—at least as far as the general public was concerned. Or in

Loraymo airfoil.

With Charles de Gaulle, discussing Loraymo airfoil.

Kennedy friendship after *Air Force One*.

Slightly out-of-focus snapshot in Moscow, as design consultant to the Soviets.

other cases, the product was appropriate but a particular manufacturer failed with his product vis-à-vis the competition because the design application was not to the prevailing aesthetic taste of the public. It is, therefore, risky for a manufacturer to either lag behind or to be too far ahead of either his competitors or the consumers' threshold of acceptance.

Various designers have created a word expressing the critical point at which one ought to stop. It is called MAYA. M-A-Y-A. Or, Most Advanced, Yet Acceptable. It is often quoted among industrial design students and teachers at universities and colleges, and especially in Germany, where industrialists often use it.

PM: *I've been meaning to ask you about individual, in many cases famous, Loewy designs. Do you think that you could tell me about some of them and point out the individual design factor that seems in retrospect the largest component in the success of the project.*

RL: That's hard to do, because what you're asking me to do is exactly counter to the design experience which views things maximally in an intregal way. However, I think in a number of cases I can think of those components that

either the public or the client most appreciated or were most salient in terms of the projects becoming most accepted, famous in some cases.

When working for NASA on the *Skylab,* I insisted that a porthole be installed on a partition so that astronauts on long missions—up to three months—could have some view of the earth. On reentry, upon debriefing, they said that the mission would probably have had to be aborted without the porthole, for psychological reasons. I also insisted on the establishment of a gravity-one-earth ambiance, that is, a ground ambiance so they could live more comfortably in more familiar surroundings while in deep space in exotic conditions.

On the Lucky Strike package, I wanted white because it looked neater, much more luminous. I then placed the red trademark on both sides—it used to be only on one side. Whichever way the pack is placed on a surface—a table, or even dropped on a sidewalk—the target is visible in any case. Many millions of packs have been sold, over twenty-five years or more. So the target, the trademark, has been visible twice as often as it would have been if it had been only on one side of the pack.

In the case of Avanti, the off-center treatment was very important, also overhead "dashboard" controls, and red illumination of the instruments at night, following an aircraft-design point of view. On the early Gestetner: ease of cleaning, and the elimination of dangerous legs. On the Concorde, with its narrow fuselage, I added a wide black color band, running from one end of the plane to the other, and placed down the center of the ceiling. This had the effect of making the fuselage look larger, an optical illusion in the last analysis, but a useful one since the side walls seem farther apart than they really are, which adds to the psychological comfort of the passengers.

Trucks. I made the seat comfortable. I made the cab comfortable too, with adjustable seats, a comfortable berth for the relief driver. All the amenities: cigarette lighters, ashtrays, a place for pad and pencil when making deliveries. The design idea was conceptual comfort, to replicate the comfort of a passenger car, plus.

For Greyhound. When the bus door is open and the passengers are ready to enter, there are some rather steep steps which may be dangerous if one is not careful. So, I placed on the door a large white disk with a large red arrow pointing to the steps, directing passengers' attention to the steps. I'm told that it has prevented untold accidents over millions of passenger miles.

PM: *How do you see your role now in the operation of your various organizations? I know you have said that you and Viola like to keep on the move, but it does seem as though you have placed yourself at the moment more in the European context than in the American. Is this true or does it just appear to be?*

RL: As the world evolves, so does my concept of how to run my own organizations. First, a designer goes where opportunities seem most interesting, and this changes from decade to decade. As I said earlier, America was almost exclusively receptive to the idea of industrial design initially. But European industrial design is fast becoming superlative in quality and innovative solutions. An industrial designer can really work anywhere now. I am told that there are now over ten thousand industrial designers in the world, with more graduating every year. Some operate in the small African nations, and I recently even heard of one established in Papua. The two companies I formed in France and England decades ago are growing fast, but the fact that they are centered there does not prevent them from working for large corporations in Switzerland, Italy, Germany, Belgium, Holland, Greece, and several African countries. Our firm in Paris is retained as a consultant by various French ministries. But it is also involved in a five-year protocol as a consultant in mass production to the Soviet Union.

Most of my personal design work is done in an isolated studio in the California desert where we spend the winter. Our organizational role has shifted somewhat to Europe, but my own view is that some of the most advanced technological research is now being done on the American West Coast. I personally travel to the site of any interesting assignment. At various times of my life I had offices or studios in Sao Paulo, Brazil; Chicago, Illinois; South Bend, Indiana; and Mexico. In fact, one of the most powerful influences on my work at a certain period of my life emerged from a business trip I took to Japan. I was on a design mission, and Viola and I were guests of the imperial family. Although we didn't actually work on any projects at that time, there I was exposed for the first time to the Japanese aesthetic by direct experience. And this in many respects has stayed with me in terms of projects I have been involved in.

PM: *Was this when you met Stuart Symington?*

RL: No, but Symington had much to do with the trip. Actually, I met Stuart Symington in 1936, when he headed a small manufacturing concern in New York. We got on well, and as a gift I designed a new type of radio for his company. His distinguished career eventually took him to Washington, and when he was appointed the first secretary of the air force, I designed his office at the Pentagon. One day much later at lunch in the Senate he said to me, "Raymond, I have to see the President (Truman). Why don't you come along; I'd like him to meet you." When we arrived at the White House, Stu introduced me and Truman said, "So you're the fellow who designed that car!" (The postwar Studebaker.) Nobody knows if it's coming or going—well, neither do I." It was Symington who, after the war, suggested and arranged that Viola and I visit Japan, a visit which, as I said, affected the rest of my career. We were received personally by Prince Takamatsu, brother of the emperor, the prime minister, and the president of the Manufacturer's Association. In fact, we were at Prime Minister Yoshida's home when a courier came to announce that General MacArthur had been recalled. (The party quickly ended.) Viola and I traveled through most of Japan impressed with the beauty, charm, and subtlety of this Eastern land which looked to the West.

When we returned to Tokyo, a surprise awaited us. The heads of Japan's industry and manufacturing were invited to attend a seminar at which my views on Japanese products in the postwar period were solicited. The main question was "Which way should we go?" When I walked into the auditorium, I found three long tables set up, with Japanese manufacturers standing behind them, each in front of his firm's main product. I was required to make short comments on short inspection, and did my best, with media covering it all. I then made a short talk, the gist of which was that Japan was known as a land that copied others' creativity. I suggested Japan's role could be greater; Japan had all the good taste, all the design sophistication it needed to become one of the world's industrial leaders. I told them that I learned much on my visit, that it would have a lasting effect on me. In Japan I rediscovered the need for clean, harmonious lines, subtle highlights and shadows, the value of understatement, beauty through simplicity.

PM: *Woulld you say that it is possible for all designers to work in a broad range? In your case, everything from the design of buses and cigarette packs, perfume bottles, whisky decanters, logotypes for firms all over the world, display units, gun systems, uniforms, hydrofoil vessels, supermarkets and department stores, oceangoing dredges, household appliances, and the like.*

RL: In theory, yes, but it is necessary to live a very long time, which is not entirely an act of choice. But I believe the designer has to choose to live internationally to not only accept but also respect the marketplace which is made up of people from every walk of life. Also, as I repeatedly say, he ought to have a background in both engineering and art history. He ought to be open to an extraordinarily broad range of influences.

PM: *Can you name some of the people with whom you've worked and described what working with them was like?*

RL: Looking back to the early days when Henry Dreyfuss, Norman Bel Geddes, Walter Teague, and Harold Van Doren, Ray Pattern, Egmont Arens, Georges Sakier, Russel Wright, and I were among the early designers who developed a profession, I can say that we proceeded with considerable naïveté. We didn't realize what the repercussions might be decades later. That a new profession was beginning around us! Aside from wanting to make a living, we were all nice fellows, pure at heart and simple enough to believe that by improving a product functionally, safely, qualitatively, and visually, we were contributing something valuable to the consumer, his sense of aesthetics, and to the country. But we also proved that good appearance was a highly salable commodity, opening the floodgates to all sorts of later unethical merchants and phony "designers" who believed that cosmetic camouflage could conceal shoddy products, increase sales, and

tickle sluggish cash registers. America became flooded with cheap, sleazy junk bought by consumers who saw gaudiness as a mark of advanced "futuristic" design.

Our small group became appalled and a bit frightened by this stampede led by fast-buck artists. We tried to stop it through articles, speeches, radio, and TV, but it only became worse, except of course, in those companies—not enough of them—managed by ethical businessmen. Today every city, town, or village is affected by it; we have entered the Neon Civilization and become a plastic world. It goes deeper than its visual manifestations, it affects moral matters; we are engaged, as astrophysicists would say, on a decaying orbit. There is a frantic race to adopt pseudotechnological attitudes or expressions, to sound scientific and avant garde; to merchandise tinsel and trash under the guise of "modernism."

PM: *What do you remember about the personalities of your early colleagues?*

RL: One thing we had in common, nearly all of us, is that we were very close friends. Most of us had our own style; we were aware of each other's work although we had our own point of view. For example, I especially liked Van Doren's work as well as his personality, and yet I can find no traces of his work in mine. Another fellow I remember is Norman Bel Geddes. Because he was such a glamorous and extroverted fellow, he made a great contribution to bringing attention to our profession. I think Dreyfuss was my closest friend; he was thorough and methodical, a good designer, but perhaps not spontaneous enough. In some ways he was right for his time, essentially safe, and therefore very successful since his projects blended in well with the contemporary taste of the public. In some ways this was also true of Walter Teague, who was, however, a first-class businesman; he was so at one with his era that even in his own time his work was studied. He was really a great fellow but entirely different from me. I sought excitement and, taking chances, I was all ready to fail in order to achieve something large.

Oh, there were others, too. Johnny Ebstein, René Labaune, Michel Buffet. I literally could name hundreds. Each of them had a contribution to make. Oh yes, Walter Chrysler, Gene Hardig. . . . I'd like to stop now because it

The scene at Le Mans with Porsche officials, Farina, Prince Berthil, and Fangio.

is a bit painful to think of these great friends and the innocence with which we approached what is now so very seriously considered.

PM: *You said before that speed has always fascinated you...*

RL: Yes, I suppose that's right. I've always been attracted to speed, and I suppose I have a secret *amour* with transportation of every sort even today. Every three or four years, you know, I still design a new car, whether or not it gets produced. Nothing gave me greater pleasure I suppose than my work on the S-1 locomotive for Pennsylvania Railroad. Urban transport now, high-speed planes, speed boats, buses, all that kind of thing, has always had its appeal. I think speed and mobility are the touchstone of our age. And as I mentioned, I like to be a participant.

PM: *How do you mean?*

RL: I like to be involved. For example, after attending Le Mans for over thirty years, hardly ever missing a year, I was made an honorary life member of the Automobile Club of l'Ouest. My friend Prince Berthil of Sweden and I regularly spent prerace time with track celebrities such as Fangio, Farina, Dr. Ferdinand Porsche, which was always an exhilarating experience.

It was rather flattering and certainly useful to know that the drivers and other race experts considered my design automotive work in the States sufficiently interesting to discuss future design trends with me, often exchanging opinions as well on racing matters. Years later the chief engineer of Porsche told me that he and his team considered the Avanti especially interesting to them due to its advanced use of aerodynamic principles.

Tom McCahill, the American automobile critic, wrote: "Raymond Loewy, Studebaker's designer and chief stylist, proved once again in 1953 that he's the guy the rest of the country's designers wish they were. Back in 1946 he inspired the industry to steal his notchback Studie designs and in 1953 he came out with a car that made the typical monsters of Detroit look as modern as Ben Hur's chariot in a stock car race. When I was in Le Mans last summer for the famous 24-hour race, I kept stumbling over Loewy

in every pit. Unlike some of our home-grown design jackasses, who never stray further from their drafting boards than the nearest saloon, here was Raymond Loewy, checking every new angle and interesting curve (automotive), the products the best car brains of Europe had produced."

I hardly think all my colleagues fit McCahill's description, but I mention Le Mans because the great majority of American automotive designers never leave Detroit, and it is my view that the racing circuit and actual contact with the drivers is a useful input in the work. There's no better test for an advanced principle than the excruciating competition of a race like Le Mans.

It may seem odd to you, but in 1963 I actually took a course of high-performance driving with Caroll Shelby, who won Le Mans several times. He had a private school at the Riverside race track in California, not far from Palm Springs. I attended for two months before receiving my certificate, although I was seventy at the time. Shelby told me I had a great racing future ahead of me!

PM: *Who built the cars you designed in France?*

RL: All the automotive-body design concepts I developed when in France were built by two talented young men whose workshop was on an abandoned farm (with curious chickens) not far from Paris: Bernard Pichon and André Parat. They also built two entirely new prototypes for Studebaker, one of them the "famous Studebaker that never was seen." This prototype was never put into actual production, but it was a running prototype which we tested at South Bend. Very few people ever saw it, but among them were those who felt that actual production might have changed Studebaker's fate.

Part of the rear section indicated a certain subtlety and slenderness of the body, but subtlety was not a popular word around Detroit in those days. Still I've always believed that consumers frequently have better taste than the businessmen give them credit for. Detroit is now discovering this, and many American automobiles are better designed these days, from the point of view of proportion and the decrease in gaudiness. Happily, a new and younger set of executives is at the corporate helm. However, in one area Detroit continues to miss. It has not entirely understood the light, compact car which it wants to make to accord with new economic realities. Such a car

To Raymond Loewy
With great admiration
George E. Mueller

To Raymond Loewy
Thanks for your contributions to the U.S. Space
Program from The Apollo XI Crew

To Raymond Loewy
With my thanks for
all you are doing to
make life more
comfortable
George E. Mueller

To Raymond Loewy
With great appreciation
George E. Mueller
Aug 69

On this and the following pages are participants in the great NASA space achievement.

Dear Mr. Loewy,

Thank you for your letter. Please excuse the delay in replying.

I remember our meeting in Huntsville very well several years ago when Skylab, then called by another name, was more imagination than reality. The porthole which was added to the spacecraft at our mutual insistence was a great asset to our operations and will be increasingly so for the last mission as they have an extensive program of earth observations to carry out. In fact, our recommendation for future spacecraft is that they have more windows and larger ones, preferably with a bubble-type configuration.

Thank you for your assistance with the Skylab design. We found it a most habitable home.

Warm regards,

Jack Lousma
Major USMC
NASA Astronaut

TO RAYMOND LOEWY WITH VERY BEST WISHES FROM THE CREW OF APOLLO 15

Dave Scott   Al Worden   Jim Irwin

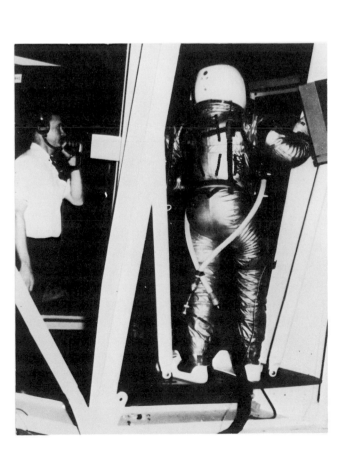

To Raymond Loewy
Thanks for your contributions to the U.S. Space Program from the Apollo XI crew

should express a new philosophy and shouldn't be a scaled-down version of the American large-car tradition.

PM: *I believe you've done some work, as well, in design architecture . . .*

RL: Oh, of course. Nearly all of my own homes, for example. Also, my design group worked on the first suburban department store in America, the Lord & Taylor store in Manhasset, Long Island. We also designed Foley's in Houston.

PM: *What about interiors?*

RL: They have personally interested me. At the moment I see a definite public trend to the elimination of walls. I think apartments and houses want larger spaces whether for parties or family living, or whatever.

PM: *That's a rather Japanese idea, isn't it?*

RL: Yes, but the Japanese do it out of necessity, since the basic space they have to work with is small.

PM: *Are you talking about flexible compartmentalization?*

RL: Yes, absolute flexibility of layout. I did that in an apartment I once had in New York. We only had walls for the bath and bedroom. Everything else was wide open, and it was a novel idea when we did it.

I think the area in which most people need the most design education is in the area of lighting. I'm against diffused lighting except in operating rooms, factories, and airport terminals. Spotlighting, pinpointing provides more variety and makes life more cheerful as well. I think most designers know this, but builders continue to provide general lighting for the masses whether in new apartments or in new homes.

PM: *You organized a museum exhibition of interior design early in your career, didn't you?*

RL: In 1933, at the invitation of the New York Metropolitan Museum of Art, I planned and designed an industrial designer's studio based on my own, which was shown

for several months. Every piece of furniture, equipment, and lighting was built to order.

The floor was black-and-white linoleum, the walls light beige, all the metallic parts were gun-metal finish without any nickel plating. Illumination was indirect incandescent. The thin trim molding on the walls was matte aluminum.

Large wall panels were compacted cork to accept thumbtacks and pins. Some evolution charts and ships' silhouettes were visible on the walls, and an automobile scale model made of clay was atop a modeling stand. Tool drawers and a storage space for clay were provided, as well as a circular pivoting platform of plastic material. (The car was a Hupmobile project.)

Storage drawers and shelves were provided on all sides, forming a continuous plastic-covered display counter. All drawers were equipped with silent roller-bearing sliding guides.

Desk tables had rounded corners, and the frames of the chairs were made of a single bent gun-metal tube. The high drafting stool at the right was equipped with a ring-type footrest. The drafting light at the left could revolve and be adjusted up or down, and the tubular lights also served as support for the desks. All ceiling fixtures were flush with the acoustic ceiling.

PM: *People talk about the personal style of a Loewy studio. What do you think they mean by it?*

RL: Well, most of my Pennsylvania Railroad and automobile designs were conceived in New York. The duplex penthouse office at 580 Fifth Avenue had two large landscaped terraces with flower beds, white gravel, bamboo armchairs, and low coffee tables. The larger one looked out on Rockefeller Center Plaza. The atmosphere was young and cheerful. We solicited everyone for contemporary design ideas, and our young secretaries not the least. We always had music playing, when this was not the fashion it is today. I maintained this type of ambiance in all our offices, in America and abroad. Our Paris office, today, is much like this one. We were work intensive, but it was personal as well.

On my birthday, the staff used to throw a party in the penthouse. Once a piano was brought up, a jazz pianist played dance tunes, and a young man came from Hobo-

In front of the house near Palm Springs, California, where the Avanti was designed in 1961.

ken to sing for us. When he left to return home, I gave him a bagful of chewing gum because we had received a full crate of sticks from a new client, Wrigley's. Our singer then told me that his ten-year-old boy had the measles and that the chewing gum would lift his spirits. Before he left to take the Hudson tubes to Hoboken, I also gave him his check for fifty dollars as agreed. Frank Sinatra then went home.

A month later he became famous at the Paramount, a few blocks away, and frantic teenagers were wrecking the place until the police arrived. We still speak of those days when we see each other in Palm Springs, where we both spend the winter.

PM: *I suppose, in terms of transportation at least, people think first of your cars and trains, but your interest in the water was always there, wasn't it?*

RL: Yes. The villa I bought in St. Tropez in 1930 was built overlooking the Mediterranean and I had a fast American

speedboat shipped there. I had already purchased a cruiser in Long Island, the *Ayrel* that I used in the early spring and in the fall, discovering the "discreet charms of New England."

In St. Tropez I enjoyed aquaplaning and underwater exploration. Compressed-air diving tanks did not exist at the time so I used a regular cast-iron diver's helmet, leaded shoes, while a crew member pumped air to the bottom. As the lubricated pump got warm in the sun, the air had an unpleasant odor; we always mixed a little Chanel Number Five with the oil. Industrial design, I thought, should apply anywhere, even while flirting with the mackerels and octopuses!

PM: *Getting back to Japanese ideas, didn't they influence the design of your villa in St. Tropez?*

RL: Yes. During our visit in Japan I purchased all sorts of exotic materials, such as black and red lacquered panels, screens, lighting fixtures made of white rice paper, large bamboo vases, silk scrolls; I also discovered bamboo cultivated square rather than cylindrical, four inches on each side. I shipped it all to St. Tropez and decided to design the house in a rather Japanese mode.

A few years later Uriane was reproduced in a Tokyo architectural magazine as an unusual example of the contemporary Japanese house. No matter that it had been built on the Riviera and designed by an American born in France! As I usually spent several summer months in the villa in St. Tropez, and naturally, needed to work, I equipped it with a studio overlooking the Mediterranean. I did some of my best known work there; this was just before World War Two broke out and our return to America. We came back to the villa after the war and found everything intact; even our devoted cook Anna happily had not suffered; we'd taken care of her financially in this dark period through the good offices of friends.

PM: *What about your other homes, the one outside Paris, for instance?*

RL: In 1933 I purchased the Manor of la Cense, built by Henry the Fourth in the sixteenth century, and it has played an important role in my life. I was born in France, of course, and still enjoy returning every year. We've always liked the area close to Paris, called Ile de France. We are surrounded by the estates of our friends, the Duke of Brissac, the Count of Pourtales, and Denis Baudoin.

The architectural style of La Cense is Renaissance, before the flamboyant era. With a large interior court enclosed on all sides, a lawn, old well, and a small pond surrounded by weeping willows, it remains an enchanting home year round. Surrounded by over a hundred acres of fields and woods and in spite of its proximity to Paris, twenty-five minutes away, there are pheasant, hare, partridge, as well as deer and wild boar. It was built by Henry for his mistress, Gabrielle d'Estrée, and has changed ownership only three times in the last three hundred years. Viola and I furnished it with a blend of antique furniture and modern paintings and sculptures by such masters as Picasso, Ernst, and Chagall.

In a former henhouse we installed a typical little French café equipped with authentic old bar, coffeemaker, marble-topped tables, café chairs, most of which we brought from Paris. In one corner stands an old seven-foot-high music box found in a nearby village which plays polkas and the popular tunes of the nineteen hundreds. When we have guests we often go to our "bistro" for coffee, a drink, and dancing. The walls are hung with posters and paintings of the period; on the bar is an authentic Edison phonograph with more than a hundred of the wax cylinders which preceded records. We still use the 1910 crank-operated phone on the wall. It works so well that one evening Dr. George Mueller, deputy administrator of NASA, called Houston's Space Center and the sound came through clearly.

We also have a country-fair shooting gallery where our guests can shoot clay pipes and iron ducks. They must pay for the shots, as at carnivals, and the proceeds go to the village's children's relief fund.

In a heavy slanted-oak-beam attic I maintain a fully equipped design studio directly connected to C.E.I., my Paris office. Two adjoining rooms are decorated in authentic Japanese style with objects, screens, and materials I shipped from Tokyo, mostly bamboo and black lacquer, Japanese actors' masks and wigs, paper lanterns, etcetera.

PM: *Very different from Tierra Caliente, your home in Palm Springs, I'd say. What are the circumstances surrounding your choice to build in the desert?*

RL: I came to love Southern California, and I wanted to own a small isolated house in the desert there. I bought some land with huge pale gray granite rocks situated in a general area of sand and cacti. Building a home for myself excited me, and the same day I bought the land, I proceeded to design it, working through most of the night, sketching, I remember, on the letterhead of the racquet club where I was staying. Using sticks, I laid out my plan to the contractor, and construction began in less than a week.

Four large rocks emerge in the pool, which itself flows into the living room. A white rug covers the room's entire floor, literally surrounding the pool. At night or when the weather is cool, two large doors slide across the pool and close off the living area. A woodburning fireplace completes the room's main features. The walls are made of pecky cypress, and everything—including the furniture—is sand colored, the exterior as well. Due to its low profile, to its color matching the rocks and the sand, the house blends with the desert and is hardly visible even as one approaches. The pool's water comes from the ten-thousand-foot Mount San Jacinto, water clear as crystal and cerulean blue to the eye. The house is full of plants and flowers, faced by a lawn and palm trees at a few chosen spots. At night, a fifteen-hundred-watt light shines from a thirty-foot, white-lacquered mast illuminating the rocks, pool, and lawn. Spotlights are located throughout the grounds, and control switches permit us to create all sorts of lighting effects, as on a stage. When the lights are off, the pool alone can be illuminated by a powerful submerged lamp, and the scene resembles a blue lagoon in a desert oasis. Due to my intense dislike of power poles and wires, I paid to have those to the front of the house installed underground. Behind the trapezoidal, pecky-cypress solarium are lemon, lime, orange, tangerine, and grapefruit trees. With a bow to the limes, daiquiris have become the house drink.

The snow-covered mountains form a backdrop. The property itself is surrounded by a fine-mesh wire screen, specially designed to prevent reptiles from entering. In fact, we've never seen one in forty years, reassuring to Viola and me since we often go barefoot.

PM: *And your home in Mexico?*

RL: Strangely, we also called it Tierra Caliente. I built a vacation home in a small village, Tetelpan, rather an artists' community in some respects: Diego Rivera and Covarrubias, for example, also lived there. I selected a hilltop site and erected a house which I actually designed in New York. It overlooked a valley leading to the Ixquatl volcano. The view across the swimming pool still means much to me. The pool, with Ixquatl reflected on its surface, was surrounded by a large lawn. Tierra Caliente's contemporary furnishings contrasted with massive sixteenth-century Spanish furniture carved out of oak and bleached by time. A large fifteenth-century mirror surrounded by a heavy rococo gilt frame dominated the absolutely plain center wall of the living room. Ceiling height was eighteen feet. The wall-to-wall plate-glass window overlooking the

pool and Ixquatl had, in its center portion, a large glassed-in area, a virtual birdcage in which dozens of colorful tropical birds and parakeets flew among flowers and palms. From the living room it seemed as though tropical birds were flying across Ixquatl.

House and Garden published a cover story about it as embodying unusual design and décor concepts, combined with the use of imaginative materials. The terrace and all rooms had pink marble floors, in the living room were wide strips of alternating black and white marble in the center. I used pale color terra-cotta for latticework and partitions; the doors were lacquered with white enamel, the living-room walls were light pinkish gray. The large sofas were covered in chocolate-brown silk, the tall drapes rough-textured and off-white. Two massive church-type candles standing in heavy sculpted, gilt holders were placed upon an early-Renaissance, bleached-oak commode eighteen feet long.

The throw rug was thick virgin white wool. Flowers everywhere, banana plants with their wide green leaves, palms and bamboo.

Bathing all, the pale blue sky of high Mexico; Tetelpan's altitude was thirty-seven hundred feet. And no mosquitoes, those dreadful beasts that spoiled for me some of the earth's most enchanting beaches, lakes, and forests.

Mexico inspired me, at least as much as Spain. I was affected by the land's rugged beauty, its blend of violence and subtlety. I went through it all with open, fascinated eyes, stopping in small taverns and tiny villages to admire and remember. Even the bullfights. I don't like bullfights as such, but feel the drama, allure, and magnificence of the spectacle, the erotic sight and sound of a fanatic crowd nearing hysteria. The parade entering the arena in the sunlight at the sound of the Virgin de la Macarena, the men advancing slowly in their *trajes de sol* is unforgettable. And then, alone on the white sand, both immobile, the black bull staring at the svelte elegance of the *torero* in his "suit of light," in deep silence. Then, suddenly, the explosive sound of the crowd as the first steps in the ballet of death are taken. Fascinated by it all, I eventually met the most famous *torero* stars of the time, Procuna, Silverio Perez, Carlos Arruza, El Soldado, and others, all subdued, reserved young men of faith and sophistication with delicate hands that Dutch masters would have loved to paint.

With David Chan, head of the Puerto Rico office.

With Tom Riedel, Pat O'Farrell (General Manager), and Mrs. O'Farrell of the London staff.

Mexico left me with a new taste for contrasts and a sense of wild basic beauty, just as Peru later gave me a sense of pure, violent colors, and Japan the value of pale tints, utter design purity, and the charm of understatement. In retrospect, I think all these experiences affected me deeply.

PM: *With your broad international design experience, could you perhaps compare two countries from a professional point of view, say the United States and the Soviet Union?*

RL: Well, one doesn't really have to talk about the American experience; it is well known and is now the general case. The problem in working in Russia has a great deal to do with spoken communications. Also, because the Russians don't know our design vocabulary and terminology, correspondence is difficult and lengthy. Their research and development is getting better, but it is slow. The one unpleasant trait in their relationship with an industrial designer has to do with what could almost be described as a pattern of changing specifications in midstream. When the design is halfway completed, they call you up and say, "No, we've decided to lengthen the wheel base." What seems to them a small change affects, as you know, the whole concept of the integral design. They don't understand that the designer has to start all over again. And they do not understand that this requires additional compensation since they tend to think of you as an employee rather than a free-lance designer, even when you are also a

retained consultant. On the positive side, among their many virtues, the Russians are very pleasant and cheerful. They pay well and promptly and have a good sense of humor and are anxious to learn. The questions they ask when I address groups are excellent and, as far as pure research goes—transcendental scientific research—the Russians are of the first rank today.

PM: *When you first went to Russia, there had been relatively few technical and cultural contacts. What was this visit like? Did it have an ideological aspect?*

RL: One day in 1962, Anastas Mikoyan, the well-known Soviet diplomat, arrived unannounced in my New York office, inviting my wife and me to visit Russia as guests of the government. We accepted and upon arrival there were given a literal red-carpet treatment, welcomed by the chief of protocol for the Soviet Union.

He assured us that we were honored guests and that his country was open to us, asked what we wished to see. We said we were interested in seeing Chekhov's house and Tolstoy's dacha, which established the tone for the entire visit. From the start we determined that the business side of the visit was to be purely professional in character, neither side would discuss matters of an ideological nature. This was kept to during our nine subsequent visits to Russia, the latest one in December, 1976.

Our interpreter was a sophisticated man, Yuri Soloviev, a designer himself. Yuri and I became close friends, and

Dune buggying in California. 1970.

he later became president of the International Society of Industrial Design as well as the director of the All Union Institute of Industrial Design and Ergonometry in Moscow.

Each time we visited Russia, Viola was invited too, and she actually handled the contractual agreements.

A Russian once told me that Karl Marx had once said that a nation has the citizens it deserves. The same fellow then said that I had been American property long enough and should now think also of Russia. Would I consider coming to Russia to help them organize the industrial design profession? I declined with thanks but recommended Yuri Soloviev for the job.

At no point in four subsequent visits did we ever solicit design work; we were simply tourists visiting many parts of a great country, making speeches to large audiences about the profession and its impact upon the everyday life of people in the West.

But in 1973 the Russian ambassador to France and the deputy minister for foreign trade, Smeliakov, came over to see me at C.E.I. in Paris. Shortly thereafter I was invited to go to Moscow with a delegation. It was suggested that I become the Soviet Union's consultant for industrial design, applied to mass production. Back in Palm Springs, Viola and I had dinner on New Year's Eve with a few friends, and by accident Henry Kissinger was there. I mentioned the Russian offer and asked his opinion. It was, said he, quite consistent with current American policy. I accepted the offer and signed our first contract. To the best of my knowledge, it was the first time such a design collaboration had been worked out with the Russians. We worked collaboratively on automobiles, farm tractors, cameras, motorcycles, refrigerators, electronic clocks and watches, and large hydrofoil vessels. Relations were good, and in 1975, in Paris, Gvishiani and I signed a new five-year protocol of cooperation as preferential industrial-design consultants.

We have always found the Russians cooperative and reliable but very hard bargainers, clever fellows especially trained in the art. I must say, however, that our years of activities, both in the States and at C.E.I. working with the U.S.S.R., were not particularly remunerative. The Russians too often changed instructions midstream, after fees had already been agreed.

PM: *Back to basics, Mr. Loewy: You once said, on the subject of simplicity, that you believed that the true answer to an industrial designer's theory of aesthetics might be— and these were your words—"It would seem that more than function itself, simplicity is the deciding factor in the aesthetic equation. One might call the process beauty through function and simplification."*

RL: I said that many years ago and I absolutely believe in that now. And yet, I always warn my colleagues that simplicity, even simplicity, *can* be carried too far. For example, a tent in the desert is perfectly adequate. It is one of the simplest forms of shelter, but who wants to live in the desert? I mean, in a tent in the desert. You see, *anything* can be carried too far.

La Cense.

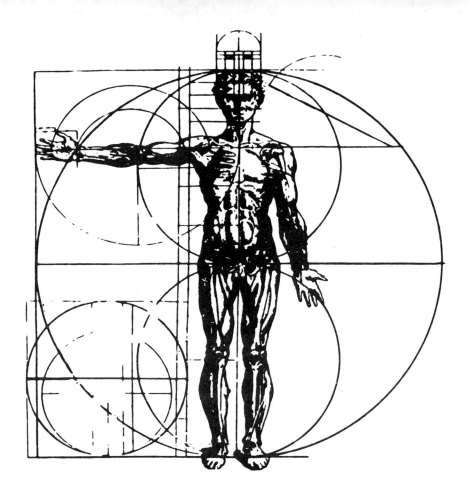

Adaptations of Leonardo's motion studies.

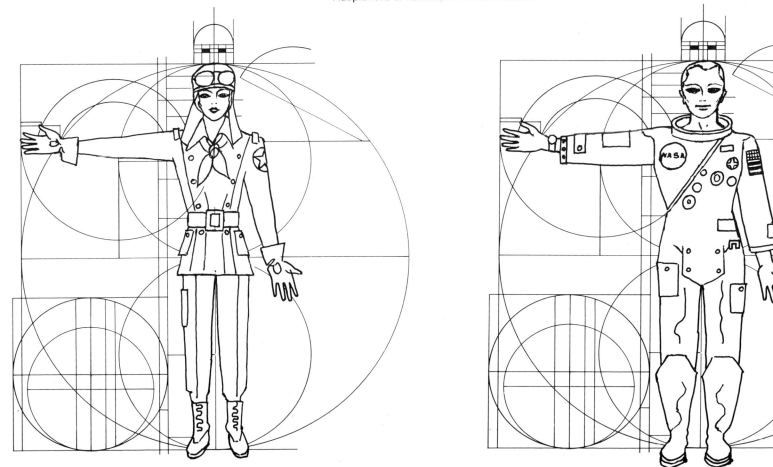

INDUSTRIAL DESIGN BY THE DECADES

*Before providing some small text to a variety of the designs and design areas which char-
acterize my work, I would like to say two things: first, it was truly impossible to show everything
within the compass of a single book, and in many cases honest men could easily differ on what
was important, representative, or influential. And in some cases it was no longer possible to
supply the illustrative material for work covering half a century. We have done our best as far
as memory and available material permitted to provide dates and sources, as well as to establish credits. Design work, even when it emerges from a specific vision, is often a collaborative
experience, and I trust the reader understands this. Second, design in industrial application takes
place within a time frame. It is culturally and historically affected and has its influence on later
periods—it obviously exists within a continuum. Therefore, somewhat in the style of a reminis-
cence, in the next few pages I will recall events, trends, movements, etc., decade by decade
so that the reader, in noting the date on a given project, can locate it, not only within my career,
but, much more important, within the time.*

THE TWENTIES

Wearing the uniform of a French army captain, I landed in New York in September 1919 with
fifty dollars in my pocket. I first visited my older brother Georges, a surgeon on the staff of the
Rockefeller Center for Medical Research in New York. I needed lodging and some kind of work,
since, unfortunately, Georges' income was not sufficient for him to help me.

I was hired as a window dresser at R. H. Macy department store and was asked to design a
display of ladies' wear. At the time Macy's windows were jam-packed with merchandise, looking like a rummage sale. On the beige carpet of the window display I casually draped a lovely
fur coat, a scarf, and a handbag, placed one yellow rose in a crystal globe, and focused a
powerful spotlight on the group. Accented by sharp highlights and deep shadows, I sought a
sophisticated look expressing quality in depth.

But the next morning Macy's top executives looked at it, and the reaction was so catastrophic
that I rushed to the personnel office and resigned before I could be fired. I made up my mind
then and there that I would never again be an employee if I could help it. My career had lasted
one day, the longest regular job I have ever held.

Fortunately, within a few days I found free-lance work doing illustrations for a fashion magazine, and, in spite of my poor English, I managed to convince Wanamaker's department store
that I could do advertising layouts as well; they were a success and I could afford to buy my

first civilian suit. Then *Vanity Fair* and *Vogue,* Saks Fifth Avenue and Bonwit Teller became clients. They liked the Parisian touch in my sketches, and I was able to rent a small office on 42nd Street overlooking Bryant Park.

It had one great advantage; over the trees I could see the office of a beautiful, young fashion editor at *Harper's Bazaar* whom I had met at a friend's dinner party. Tookie was slender, tanned, Californian, witty, and very chic; she became my friend, and the next few years were a delight. She knew all of New York's "in" people, many of whom we met at the round table of the Algonquin, where Tookie and I often had lunch. Tookie had been raised by a Chinese nurse in Berkeley, and we both spent exotic evenings in Chinatown, mostly at a tiny Chinese theater where the colors and slow motions were exciting. Tookie helped me to learn English, and when I later bought my first boat, she introduced me to the summer charms of Nantucket, Cape Cod, and the steamed clams, lobsters, and corn of Maine's sunny beaches. My early encounter with sophisticated New York and the half-tone subtleties of New England are memories I owe her. Life was very pleasant . . . until 1929 and the Wall Street disaster.

The Great Depression shook America and my life; I decided that my fashion sketches and Chinatown days were over and that more serious challenges should be faced.

THE THIRTIES

The thirties essentially saw the beginning of real industrial design in America. Having ended my career as a fashion illustrator, with vitality and conviction I pursued a new goal for life: industrial design. I gave it nearly fifty years of effort throughout depression, war, and prosperity.

Shocked by the fact that most preeminent engineers, executive geniuses, and financial titans seemed to live in an aesthetic vacuum, I had enough effrontery—or recklessness, perhaps—to believe that I could add something to the field.

The beginnings were hard; people were rough, antagonistic, often resentful. Unlike my early colleagues, most of them born American and from well-off families with enough resources to keep them going in difficult times, I was much more on my own, with depleted funds in an impoverished America. My French accent was no help, now that I was no longer in the fashion world. I remember long, gloomy winter nights in cheap hotels in Cicero, Illinois, feeling sick, tired, and discouraged after even longer days trying unsuccessfully to sell industrial design to minor, but tough midwestern manufacturers. Dismal memories: cold beds, cold meals, cold rain, and plenty of aspirin.

In 1933, the new president bucked-up the nation: Franklin D. Roosevelt uttered his famous phrase over the radio: "The only thing we have to fear is fear itself." America, with the rest of the world, began its long struggle out of depression. A lack of imaginative products and advanced manufacturing was a form of fear that had contributed to the general economic decline. The problems all over the world were, of course, much larger, but one of the solutions was to create consumer demand. Eventually, a few industrial design pioneers were able to make some business leaders aware that this lack of vision and industrial timidity was foreign to the spirit of adventure that had made America a leading nation. And could again.

Success finally came when we were able to convince some creative men that good appearance was a salable commodity, that it often cut costs, enhanced a product's prestige, raised corporate profits, benefited the customer, and increased employment. The worst was far from over when World War II started, but things had improved from the dark days of the early thirties.

The pages that follow describe some early examples that have now become part of a profession's saga. The record is widely quoted and entered in libraries and museums internationally. On the occasion of the U.S. Bicentennial it was officially noted that the American standard of living had gone up, in no small part due to the contribution of industrial design. Many other nations in Europe and Asia have successfully applied the American model to their own problems —Japan conspicuous among them.

THE FORTIES

During the war years, all of us who had started out in industrial design lent ourselves mightily to the war effort. One of the more amusing contributions, in retrospect, was a swivel cardboard lipstick container that sold in the millions. The task was to create a lipstick holder without using any critical war metals. The commission came from Washington and the goal was "to help maintain the morale of the American woman." I like to think that in creating the container I did something for American men as well! With the Japanese surrender in 1945, the opportunity arose to originate an industrial model to set the stage for a prosperous peace. Much happened quickly and the times were of use: depression and war were behind us, Americans turned their minds to the good life, as they saw it, with a heavy emphasis on the demand for better consumer products. Commerce and transportation geared itself to the new order and industrial designers came to the fore, called in by America's largest corporations for assistance. My firm was retained by General Motors' Frigidaire Division, Greyhound Bus System, Shell, BP, Coca-Cola, and United Airlines. Many of these clients stayed with us for decades.

In this period I designed and built several experimental personal cars which influenced later designs for automotive manufacturers in Detroit. I traveled throughout the forties, especially to

Latin America and to Mexico, where I built a winter home. Viola Erickson and I were married in 1948, and on one of our trips to France we were honored at the Hotel de Ville as honorary citizens of Paris. I took special pleasure in this since I remembered leaving France in 1919, virtually penniless and with no clear idea as to my future. Viola was even made honorary sergeant of my old World War I regiment; this took place in front of assembled troops at the Mount Valérien in Paris. These honors added nothing to the world of design, but great gusto to our lives!

THE FIFTIES

By this time the principles of industrial design and the efficacy of a concern for both aesthetics and function were well established. No firm of any size or sophistication in the United States was unaware of them or could do without them. The field blossomed, and although, as in all fields, inferior work was sometimes done, in general progress was made and it benefited all parts of the society.

Firms like Shell and Nabisco moved on a fast track, understanding the need for a total design sense, not simply the industrial design of individual products. We in the Loewy organization developed for Shell the new "ranch" station, which became a much imitated or developed-upon prototype; it also marked the beginning of our twenty-five-year-long consultancy with them both in the United States and with Shell International. A breakthrough in Nabisco packaging had similar long-term results.

In this period I was invited by the Japanese government to visit that country—a trip sponsored as well by the American government, which had an understandable interest in Japan's peaceful development. We had numerous meetings with the assembled leaders of Japanese industry and finance to help establish a blueprint for Japan's postwar industry.

The fifties saw our range broaden as a firm. We worked on projects as varied as the French helicopter Alouette, a BMW competition car, an ESV (experimental safety car) for the U.S. Department of Transportation, cargo ships, and nuclear trans-Atlantic vessels. Modern transportation continued to fascinate me, and during this time we also took long cruises to distant spots on the eighty-five-foot yacht *Loraymo*, built by Camper Nicholson, with a design studio aboard.

THE SIXTIES

With the battle well won on the principles of industrial design and with a large number of practitioners of varied talents in the field, the sixties presented opportunities for me and my organization to become involved with larger projects, often of a governmental and/or scientific nature.

In this remarkable decade I spent a great deal of time in Washington, especially with the Kennedys, and I received the commission to design the exterior markings of *Air Force One*, as well as the interior appointments for the president's personal use. I often visited him in the White House, where we worked on a plan which was to have national scope: the application of standards and concepts of industrial design to some of the important issues of the period, including highway congestion, urban blight, and redesign of government buildings. A team of his close collaborators was formed to help me get started, and Arthur Schlesinger, Jr., was assigned as my permanent contact at the White House. At another time the president also recommended me as a consultant to Jim Webb at NASA; this later led to the most important assignment of my life.

My involvement with the Kennedys was extensive; my wife and I spent weekends with them on the Cape and became close friends with Rose Kennedy and Ted. The president's death was a great personal loss to us, and I gladly became a member of the small group planning the Kennedy Memorial. I was later asked and was happy to accept the offer of his widow, Jacqueline, to design the Kennedy memorial postage stamp.

I was retained by NASA as a habitability consultant (to be consulted on any matter relating to the psycho-physiological safety and comfort of a crew functioning in conditions of zero gravity) for the *Saturn-Apollo Skylab* projects. From 1967 on, I worked on the Apollo Application Program, including *Skylab*'s interior, as well as on the development of the first concepts for the recuperable earth-orbital shuttle.

Although working for the government was quite different from working for private industry, the professional issues facing me were identical. The style of work and the personalities were different because the content was not competitive in exactly the same way; after all, we were not making products for the marketplace. At the outset I encountered resistance from NASA engineers and astronauts concerning the philosophy of design for lengthy space missions. At NASA headquarters in Washington, during a plenary conference organized by the deputy administrator for manned space flights, the NASA leadership opted for my views and I was chosen to address two hundred engineers, scientists, space-medicine physicians, and astronauts to explain the concepts I believed in.

The *Skylab* missions were successful, as we all well know, some of three month's duration in space. As far as my part in the project, I am happy to say that every single crew member returned safely, both physically and mentally fit. The numerous official congratulations and commendations meant much to both me and to the field of industrial design.

The sixties did not see me or the Loewy organization give up our commercial connections. We continued to work for our traditional clients and took on new ones such as Jersey Standard, for whom we developed the Exxon (née Esso) trademark, and Studebaker—the Avanti also hails from this period.

Viewed as a whole, the sixties represents the blossoming of our involvement with government and scientific and cultural institutions. I frequently lectured at Harvard and MIT and, in collaboration with Canadian architects, planned and designed the Sir Etienne Cartier Center for the Performing Arts in Montreal.

THE SEVENTIES

This period embodied a major shift in my preoccupations, although not in my life. Viola—now deeply involved in our business offices all over the world—and I transferred the center of our design activities to Europe. Much of our effort has been to accelerate the healthy progress of our organizations in London and Paris. We sold the American company but retain the name worldwide of Raymond Loewy International. We have recently started another new firm in Freiburg, Switzerland, and we are exploring the possibilities of opening a compact think tank on the West Coast to supply our European offices with California's spirit of advanced design technology and innovative thinking.

In 1975, the Smithsonian Institution in Washington opened a four-month exposition, gathering together some of my work which, over more than four decades, had become associated with the ideas of industrial design. While working closely with the Smithsonian, I was provided with an opportunity to reasses the past and to look to the future.

This decade also saw the signing of a protocol of cooperation with the Soviet Union. We became industrial design consultants for their major industries and, for a while, I spent as much time in Russia as in America or Europe. For Shell International, we developed an effective system of self-service, an improved trademark, new prototype service stations, and a design upgrading of their existing 130,000 outlets. Governments made greater use of our services, and we were engaged to design advanced railroad equipment for both the French and Dutch governments. We became long-term consultants to Total, France's largest oil company, and we undertook the interior design planning for the French Concorde. De Dietrich, France's large manufacturer of quality household appliances, commissioned us to provide them with a comprehensive design policy, and we were brought in as consultants for the new subway system in Teheran. In England our clients discovered our innovative approach to all areas of packaging and graphics. All in all, a busy decade, as my associates and I move into the eighties.

Issued in 1928, this U.S. Design Patent presented novel features that manufacturers used only decades later: slanted windshield (they used to be vertical), streamlined body (they were "boxes on wheels"), down-slanted hood (they were horizontal), streamline tucked-under rear end (they were boxy), rounded rea roof shape (they were square), shaped fenders, and horizontal hood louvers (they were vertical). The enti car had a dynamic look of motion; the automobiles the era appeared static. ▶

LOEWY 28

RENAULT

A FRENCH CAR THAT INTERPRETS IN MECHANICAL TERMS, A BRILLIANT HERITAGE OF ART AND CULTURE

A TWENTIETH CENTURY EXPRESSION OF THE FRENCH CIVILIZATION
▼
FULLY EQUIPPED RENAULTS PRICED FROM $1,950 TO $12,000 INCLUDING THE TAX.
▼
RENAULT-719 FIFTH AVENUE NEW YORK-SERVICE AND PARTS-776 ELEVENTH AVENUE.

This ad, designed in 1928 while I worked in the New York world of high fashion, reflected fashion's concern for elegance. It is what the Renault people were then looking for, and the drawing and typography were quite new to Madison Avenue advertising agencies.

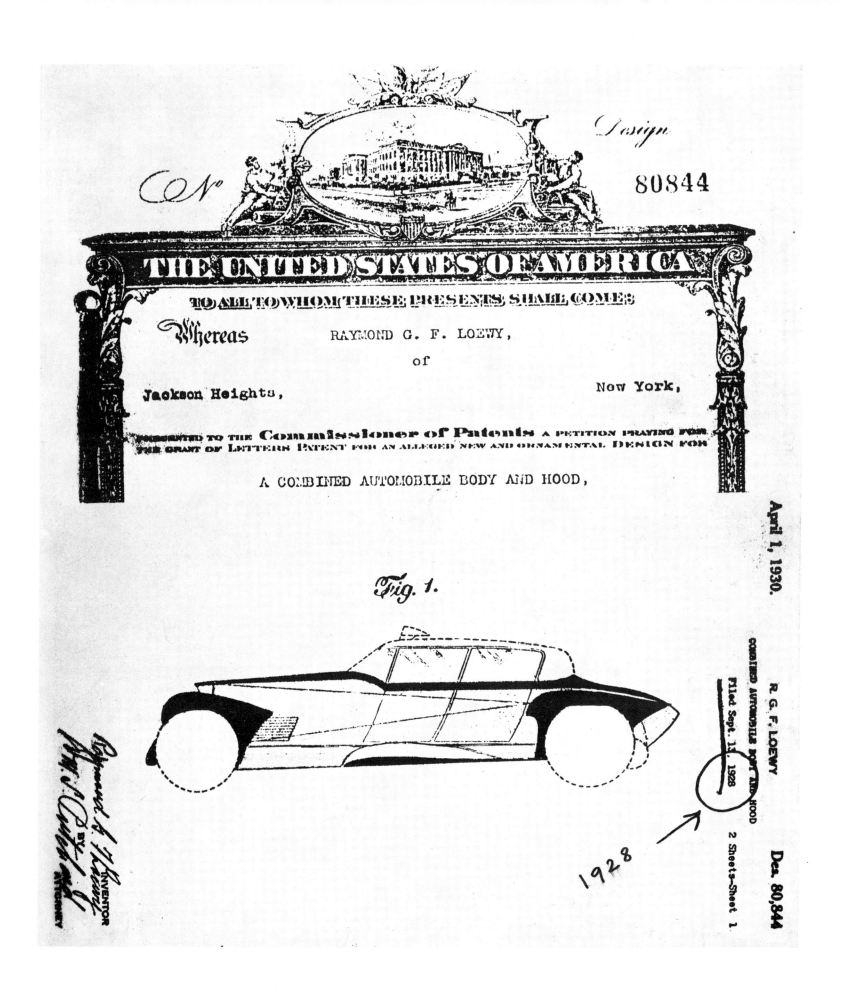

neiman marcus
1923

Young Stanley Marcus, whom I met in New York in 1923, asked me if I would design a new trademark for his great Neiman-Marcus store in Dallas. I chose a simple design in black, beige, and gold, and it was used for many years.

Advertising liquor accessories during Prohibition was touchy, but this *New Yorker* ad managed to do so tastefully, and it attracted many favorable comments. The couple doing a tango and the girl at the bottom established the mood of the time. Captions were rather witty. ▶

CHEV-ALIER OF THE LEGION OF HONOR The cross of the famous French legion has just been conferred on Raymond Loewy, New York artist closely identified with the modern movement in decorative art.
(Underwood)

Distinguished English arrivals that present the case when a case in the hand is worth two, two miles out, or on top of a truck, or well, you finish it, bottoms up. P.S.— An exclusive story, from London to Saks-Fifth Avenue.

A capacious case for that precocious Sundae School class of Elinor Wylie's (see New Yorker, Feb. 19.) English cowhide case, with two metal quart flasks, shaker, four cups (right). 48.50

A dandy brief case surprise! Of fine London cowhide, it holds three glass flasks, and the case has double locks. 34.50. (Above)

Always invited to go on the pleasantest voyages — traveling case with everything you want, i. e., shaker, vacuum ice bottle, 2 nickel and glass bottles, 6 cups. Cowhide case, suede lined. 85.00.

The popular big brother to powder compact is the refreshment compact — a folding shaker, 4 cups, squeezer, sugar box; all fit neatly into a cowhide case. 12.50.

We furnish the case—you furnish the fun! And what a case! Handsome black cowhide, sumptuously outfitted in sterling silver —two quart flasks, 8 cups, quart shaker. 375.00—and worth it!

LEATHER GOODS DEPARTMENT—STREET FLOOR

SAKS ~ FIFTH AVENUE
FORTY-NINTH to FIFTIETH STREET

The executives of Bonwit Teller's department store wanted to issue a magazine occasionally, one that would reflect the store's elegance and its New York style.

The back cover, printed in matte gold.

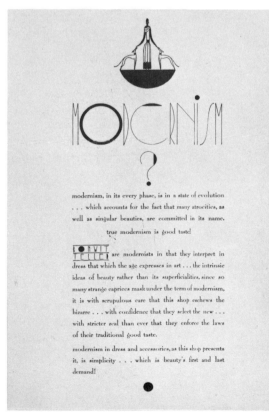

The first page, a definition of *modernism*, a word that may sound quaint or awkward today, but that then represented the avant-garde spirit in literature, music, and the arts.

Gestetner

1929

The Gestetner duplicating machine, 1929, is generally considered the first American example of industrial design before industrial design was understood as a conscious activity. The design notions introduced in it over five decades ago have survived, and the Gestetner is a classic example of what the profession can contribute to a manufacturer's lasting success. It deserves attention for another reason: it points out the early differences between a straight engineering approach and the designer's attitude when faced with the same problem—in this particular case note the four protruding tubular supports. As a consumer-conscious designer, I detected the inherent hazards of the four protruding legs in a busy office. While my client, Sigmund Gestetner, seemed hesitant about giving me the redesign assignment, I quickly sketched a stenographer tripping over a leg, paper flying everywhere. This sold him, and I got the job. During the early days of industrial design, I often used a rough-sketch device to promote what we could do for a client.

I found this yellowed snapshot of Sigmund Gestetner looking at the clay model of his new machine for the first time. When it went on the market, it sold essentially unchanged for forty years. I often kidded Sigmund about the fact that an exceptionally successful design can make a fortune for the client and put the designer nearly out of business, waiting forty years for the next assignment. This was far from the case, however, since we designed many other products for Gestetner over the years, and now, fifty years later, C.E.I. is developing other designs for Gestetner.

The first model change, 1929.

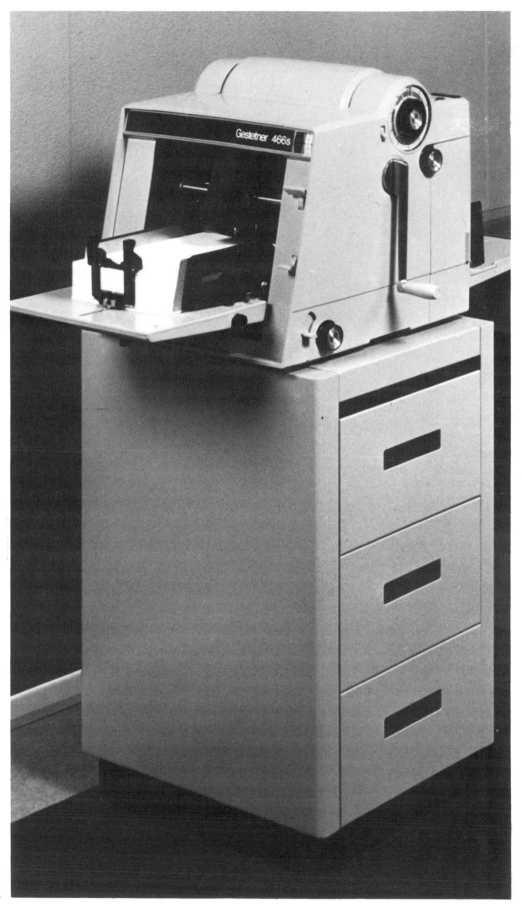

1968.

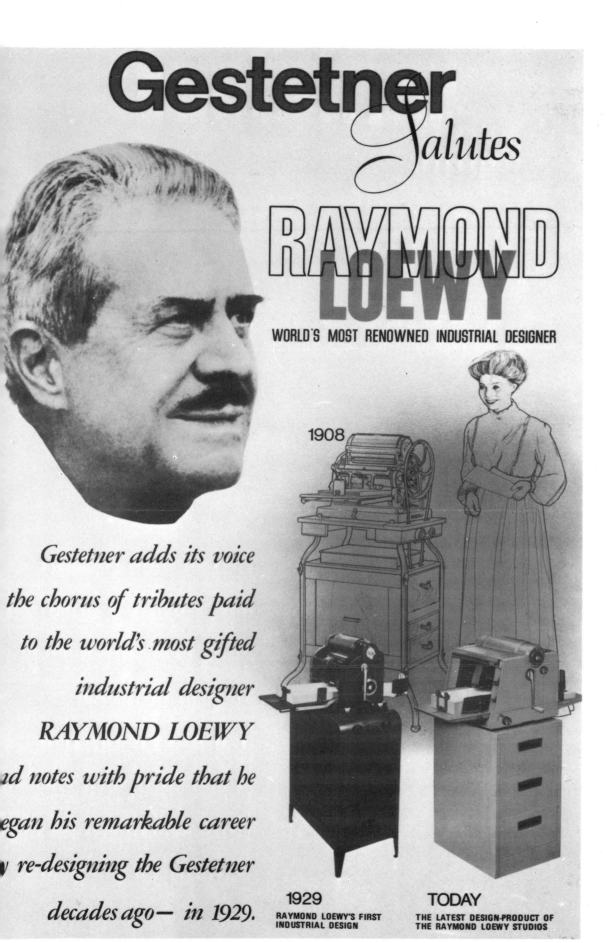

Gestetner *Salutes*

RAYMOND LOEWY

WORLD'S MOST RENOWNED INDUSTRIAL DESIGNER

Gestetner adds its voice the chorus of tributes paid to the world's most gifted industrial designer RAYMOND LOEWY d notes with pride that he egan his remarkable career y re-designing the Gestetner decades ago— in 1929.

1908

1929
RAYMOND LOEWY'S FIRST
INDUSTRIAL DESIGN

TODAY
THE LATEST DESIGN-PRODUCT OF
THE RAYMOND LOEWY STUDIOS

1978.

1976, Gestetner printed and displayed this large color poster in their offices around the world.

Hupmobile

In the early thirties, most cars looked something like this sketch.

The new Hupmobile was a radical departure and one of the first models honored by the Classic Car Society of America.

A glance at this sketch indicates that the car was more than a different artistic idea. It was a new aesthetic concept expressing a feeling of simplicity. The rendering, with its interplay of highlights and shadows, its fleetness and motion, was itself a fresh concept, which in turn had a definite effect upon the automotive industry's advertising.

The Hupmobile case is typical of an early concern not only for the product itself, but for all elements associated with total design quality. Aside from the design elements already mentioned, I took into consideration the advertising catalogues and operational manuals, the showrooms' décor and lighting, and the company's letterhead. I even convinced the firm's president that his office and some of the supporting staff were superannuated.

The 1932 Hupmobile was not accessible to the average person; I turned to mass transportation with this design for the Greyhound bus in 1946. ▶

1932

Tierra Caliente, Palm Springs.

The built-in spare wheel reduced turbulence and the large rear windows improved rear visibility. The curved rear end gave the body a flowing appearance. The rear view illustrates again the tapered shape of the entire shell, away from the boxy look and toward more naturally harmonious lines. Air resistance was reduced at high speeds. The rear lights became integrated with the whole.

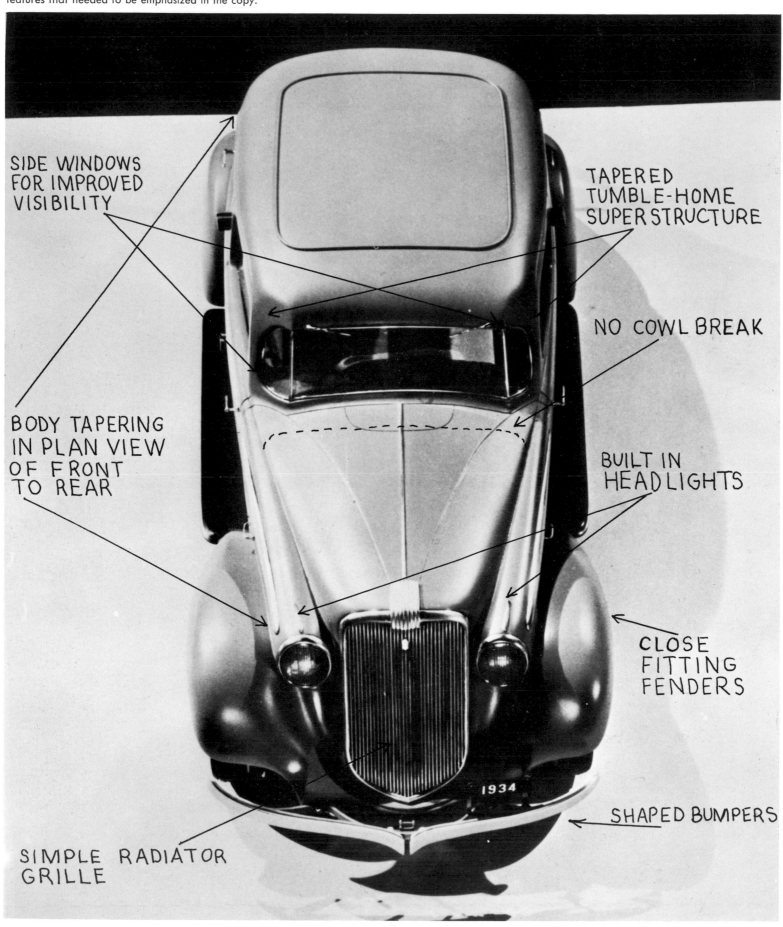

SIDE WINDOWS FOR IMPROVED VISIBILITY

TAPERED TUMBLE-HOME SUPERSTRUCTURE

NO COWL BREAK

BODY TAPERING IN PLAN VIEW OF FRONT TO REAR

BUILT IN HEADLIGHTS

CLOSE FITTING FENDERS

1934

SHAPED BUMPERS

SIMPLE RADIATOR GRILLE

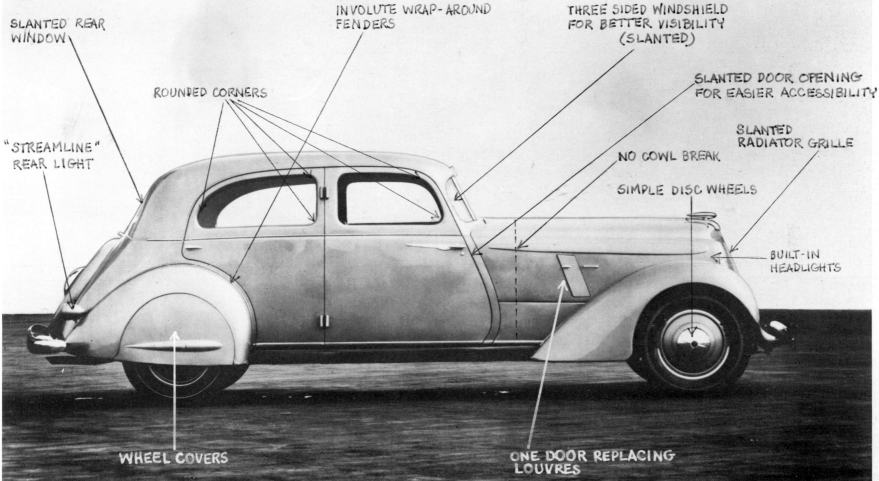

SLANTED REAR WINDOW

INVOLUTE WRAP-AROUND FENDERS

THREE SIDED WINDSHIELD FOR BETTER VISIBILITY (SLANTED)

ROUNDED CORNERS

SLANTED DOOR OPENING FOR EASIER ACCESSIBILITY

"STREAMLINE" REAR LIGHT

SLANTED RADIATOR GRILLE

NO COWL BREAK

SIMPLE DISC WHEELS

BUILT-IN HEADLIGHTS

WHEEL COVERS

ONE DOOR REPLACING LOUVRES

Many of these features, now taken for granted, were new forty-five years ago. For instance, the built-in headlights, the suppression of the unnecessary cowl break, the tapered "tumble-home" superstructure, and close-fitting fenders.

The slanted windshield and rear end, combined with the tapered sides, gave the car a look of speed. The involuted curve of the fenders, the slanted radiator grille, the built-in headlights, the soft-cornered windows, the large single ventilating louvers on each side of the hood, the slightly dipping hood effect, the simple wheel cover, all contributed to supply the Hupmobile with a touch of elegance new to the Detroit of that period.

The 1932 Hupmobile won first prize in the major auto shows in which it was entered, including those in Paris, Monte Carlo, Nice, Cannes, and Le Touquet. The manufacturer widely mentioned these awards in magazines, newspapers, and on radio; and of course, in all ads and sales literature for years.

The existing hub cap, made of several assembled components, including a colored enamel escutcheon, was expensive and rather unattractive. I designed another, stamped in one piece and chrome plated, which was preferred. The cost of each was cut by 50 percent. Due to its design, it sparkled when in motion, giving the entire car a lively look. There is currently a Hupmobile fans' association, and it meets frequently. In 1977, I attended a get-together in which dozens of forty-five-year-old cars appeared, shiny and running well.

While designing the Hupmobile I encountered resistance from a management skeptical of the new ideas I recommended. Were these ideas sound, feasible, and of a permanent character? The executives were hesitant. I decided that the only way to obtain their approval was to build a car by hand, starting from a standard chassis, and to show them that I was right. It took several months to construct the car, but when the engineers and executives saw the innovations in three dimensions, they agreed, and my task became easy. This exercise cost me personally over twenty thousand dollars (in 1932!), but it was worth it.

The body was black, the interior beige leather. Contrary to convertibles of the era, the top, when folded, was flush with the belt line.

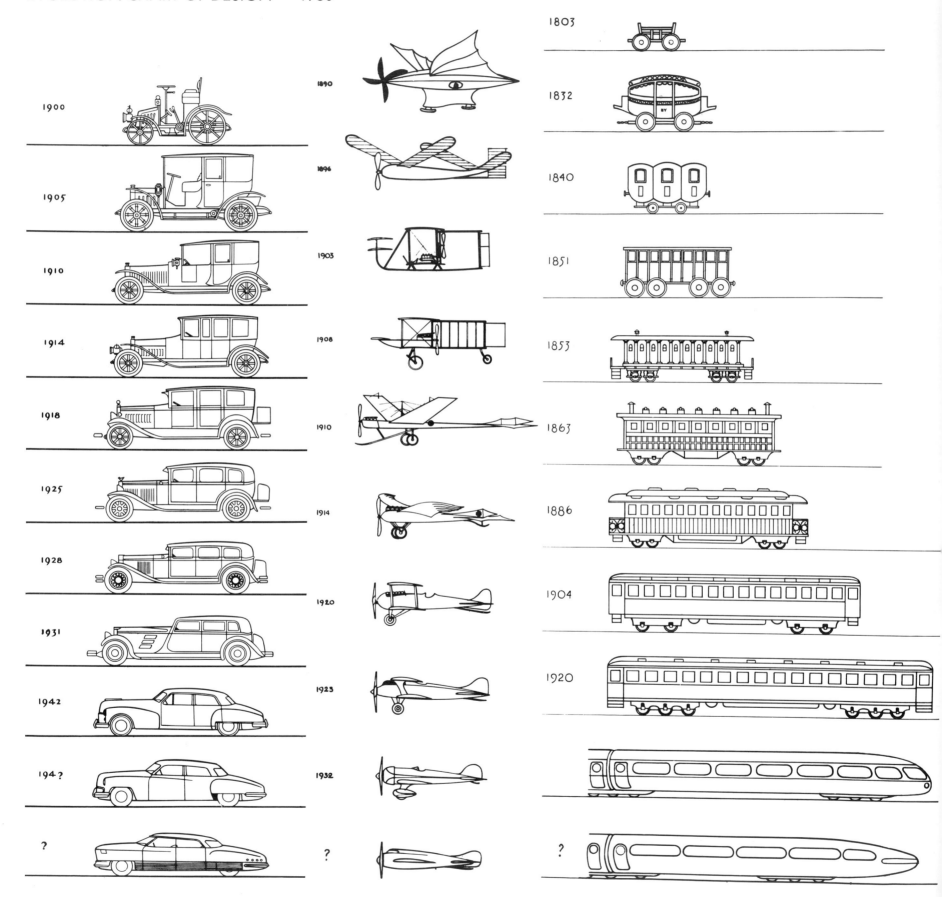

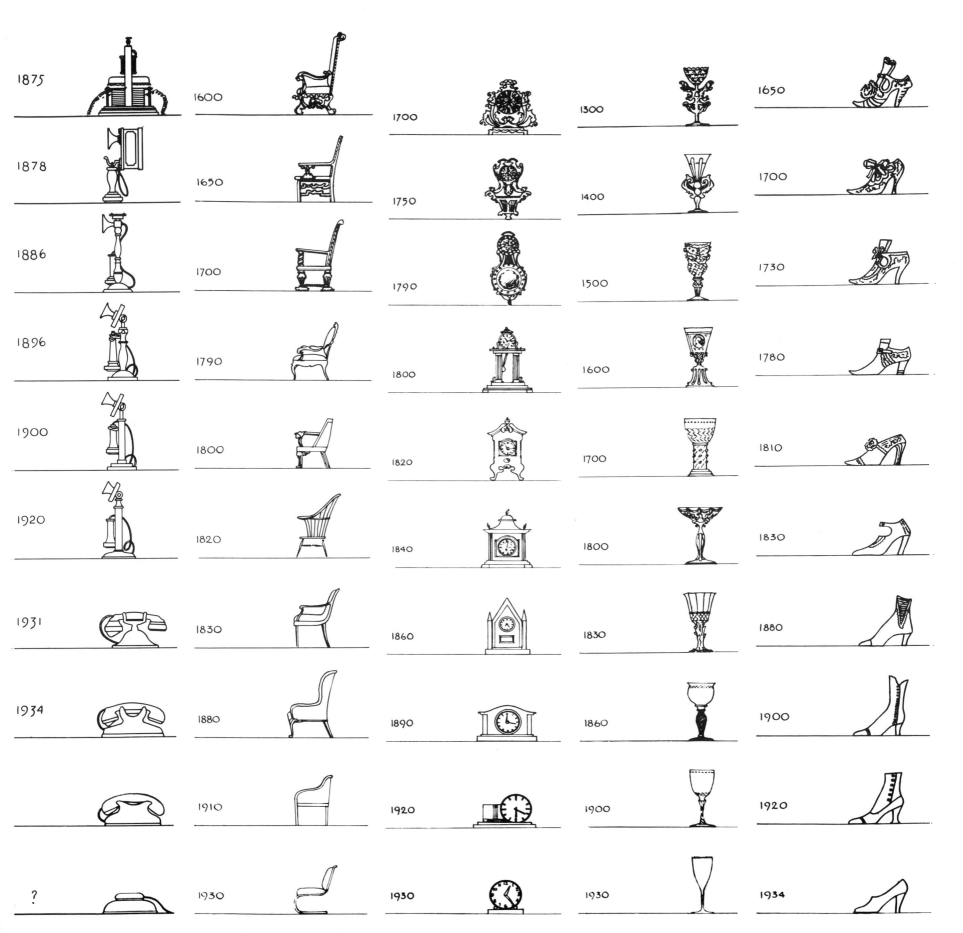

1875
1600
1700
1300
1650

1878
1650
1750
1400
1700

1886
1700
1790
1500
1730

1896
1790
1800
1600
1780

1900
1800
1820
1700
1810

1920
1820
1840
1800
1830

1931
1830
1860
1830
1880

1934
1880
1890
1860
1900

1910
1920
1900
1920

?
1930
1930
1930
1934

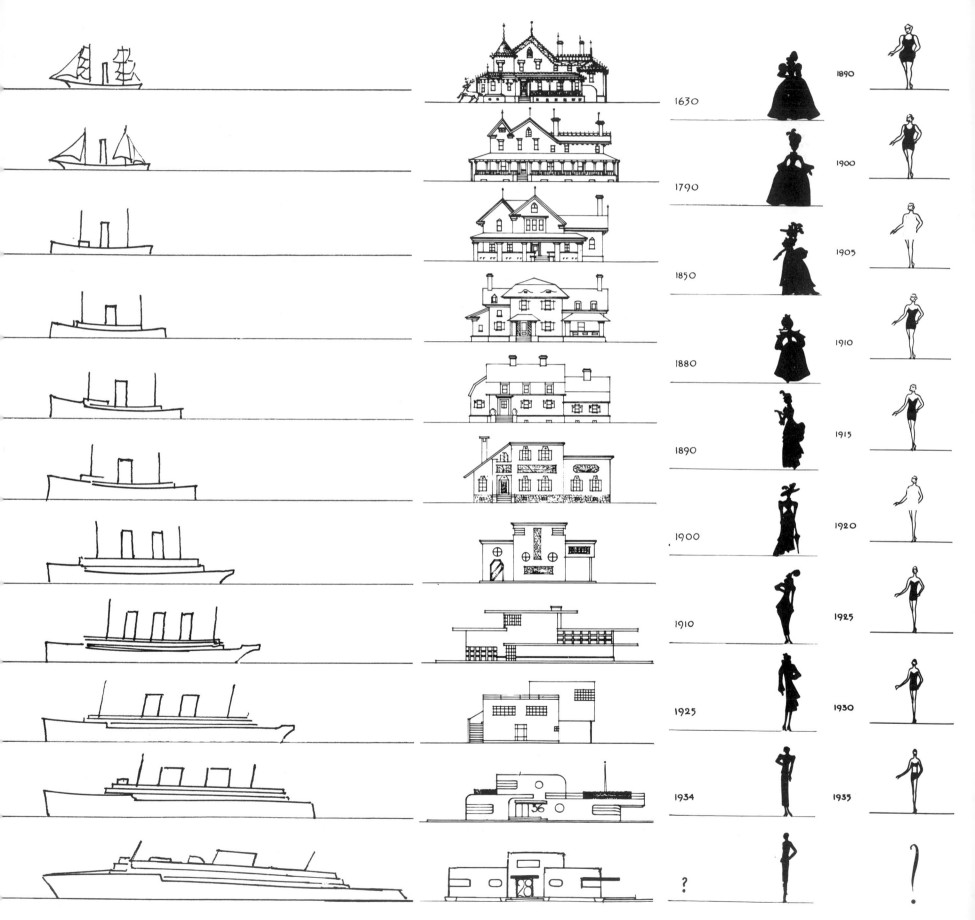

1630

1790

1850

1880

1890

1900

1910

1925

1934

?

1890

1900

1905

1910

1915

1920

1925

1930

1935

?

General silhouette of a steam locomotive of the forties.

My decades as designer for the Pennsylvania Railroad seem in retrospect almost a career in itself. It started in 1932 when the firm's president asked me, as a way of getting rid of the young "Frenchy" (as he called me), if I could design a trash can for Penn Station in New York. I did, and it was simple to use, easy to clean, and good looking. Besides, it was cheap to fabricate and—silent.

The president was intriguing, a handsome, white-haired giant of a man who sat on a high-back chair, like Mussolini behind his desk, in a huge black and brown office matching his dark clothes and black necktie. I knew that his engineering staff was planning an experimental triplex locomotive. Without any commission, or even discussion of the project, I made my own rendering of an advanced design, showed it with trepidation to him. The reaction was unexpectedly encouraging. Although the engine was never built, the presentation achieved its effect. The meeting led to a long series of major Pennsylvania Railroad assignments.

THE PENNSYLVANIA RAILROAD

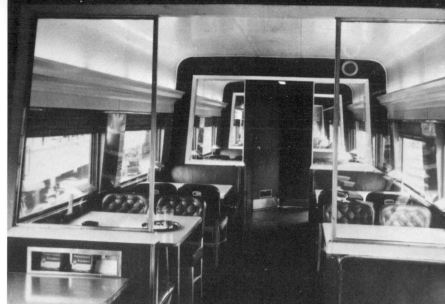

Lounges on the Pennsy's Broadway Limited, 1935–1940. Note tapered sides.

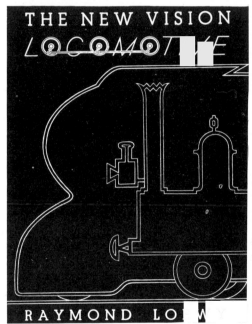

A book I wrote on the international aesthetics of locomotives, 1937.

Another "blue sky" rendering of an imaginary locomotive. I knew it would never be considered, but repeated exposure of railroad people to this kind of advanced, unexpected stuff had a beneficial effect. It gradually conditioned them to accept more progressive designs.

When the Pennsylvania Railroad asked me to prepare a design for the K4S steam locomotive, I suggested to my friend, chief engineer Fred Hankins, that we should pay attention to aerodynamics and try for reasons of safety to develop a system to lift the smoke over the top of the engineer's cab. K4S was not to be just another locomotive; I wished to make some basic improvements, and he agreed to help me.

In order to find out exactly what happened to the airstream at speeds often above one hundred miles per hour, I suggested that I be allowed to ride in the cab of a speeding locomotive. Hankins arranged it, and I was given authorization to ride a stretch between Chicago and Fort Wayne, where the tracks were on a straight line. It was late fall and I wore a warm sweater, a fur-lined anorak, tweed cap, goggles, and heavy gloves. During the ride I held a short stick, like a bandleader's baton, to which a white ribbon was securely attached. With a good grip on the forward handrail, I moved the stick in the other hand in all directions and watched the ribbon behave in the airstream: a pattern became discernible.

This was most instructive, and when I got off at an intermediate stop, I had acquired plenty of interesting data about air turbulence at high speeds . . . and how to catch a cold in a few minutes! I rejoined the engineer and fireman in the cab up to Fort Wayne and made another discovery about locomotives—there are no toilets. I decided that the new locomotive would have a toilet . . . and eventually I won.

In New York we started a scale clay model of a new steam locomotive. I had a hunch that a flat plane placed forward atop the locomotive, immediately behind the smokestack and flush with it, might help. By giving this plane a slightly streamlined configuration, like an airplane wing, it would create at high speed a depression over the engine and lift the smoke over the cab, improving forward vision of the track.

Hankins went along with the idea and it worked; the K4S was noticed and studied by railroad-engineering staffs in foreign countries, and many of its features were incorporated in future locomotive design.

The GG-1 electric locomotive was a new departure in railroad technology. Formerly the engine's shell was an assemblage of shaped steel plates riveted together.

I suggested to the Pennsylvania Railroad executives that the plates instead be "butt-welded" into one single unit, assembled on the ground, then lifted up and lowered over the locomotive's chassis, the way cars were mass assembled in Detroit. This system proved to be more efficient, and the smooth steel—free from thousands of protruding rivets—was easier to keep clean. The body was black with horizontal gold bands curved downward at both ends for a reason: the GG-1 was so silent that working track-maintenance crews met with accidents in the early days of GG-1's operation; on one occasion a crew did not hear nor see the locomotive arriving at one hundred miles per hour, with sad results. I then recommended that a bright

reflecting cluster of gold lines might make GG-1 more visible, which proved to be the case. The illustration shows the high degree of light reflection of the gold-stripe cluster on the black shell.

A ceremony was held in Washington in May 1976 by the Friends of the GG-1, a group of railroad aficionados who had completely renovated one of the first GG-1s still in service; the meeting was attended by hundreds of members and the president of Amtrak. I was their guest of honor and made a short speech, received a diploma as an honorary life member of the American Railroads Society, then returned to Penn Station in New York in the private car of a former American president: all in all a lovely day in the lovely world of old railroading.

When war was declared and young Americans left for overseas battlefields, I suggested that the railroad people who were contributing so much be present as well, at least in a symbolic way. A flag and three huge black-and-white portrait photographs of an engineer, a fireman, and a conductor gave Penn Station a note of confidence, power, and dignity, and provided a fond farewell to those who were leaving friends and family behind. At night the flag and faces—so typically American—were illuminated by powerful projectors.

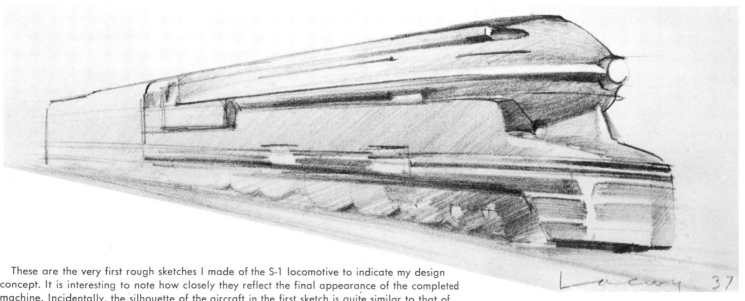

Color fold-out: 1937 T-1. ▶▶

These are the very first rough sketches I made of the S-1 locomotive to indicate my design concept. It is interesting to note how closely they reflect the final appearance of the completed machine. Incidentally, the silhouette of the aircraft in the first sketch is quite similar to that of the Concorde, built four decades later.

1962 Hydrofoil.

This is the first sketch I submitted for a staggered-level rear observation car. It provided a clear view of the tracks fast disappearing in the distance, quite a thrill for riders when first introduced. A glassed-in superstructure equipped with armchairs made it possible to watch the sinuous motions of the long train taking curves at high speed along the beautiful shores of the Hudson River, an unforgettable experience especially when the leaves turned in fall. Many of these cars were built and operated, adding interest and, some said, romance to American railroading.

An early GG-1 at the Wilmington, Delaware, Pennsy construction yard, spring 1934. More than one hundred were built, and they were widely considered the best high-power, high-speed electric locomotives ever produced in America. John V. B. Duer was the engineer in charge, a remarkable man who left his mark upon American railroading. These "motors" are still in operation, after forty-seven years and billions of miles. Here a completed GG-1 shell is lowered over its chassis. The shell of the next GG-1, made ready, can be seen in the background.

The tough yet graceful T-1 engines weigh about a million pounds. They were built to avoid doubleheading heavy passenger trains. They could haul 880 tons at one hundred miles per hour and take care of the tremendously increased wartime traffic; fortunately they had been ordered seventeen months before Pearl Harbor. The boiler jacket was designed to be flush with the top of the cab. The space gained provided ample room for all the auxiliary devices, steel excrescences that formerly protruded above the engine unit.

The forward pilot, made of heavy steel plate for protection in case of collision at high speed, contains major equipment. The inter-cooler, bell, air pump, and coupler were all incorporated within the pilot to form a flowing, smooth surface. A high running-board skirting, with handrail running the whole length of the engine on both sides, provided access to all units from the pilot to cab. Each locomotive made the 713-mile trip from New York to Chicago with only one stop for fuel. Fifty-three T-1s were built, and they proved remarkably trouble-free and reliable. (See also full color fold-out.)

The 6,000-horsepower streamlined S-1 locomotive and its smoke-deflecting device, an improved version of the K4S. When the clay scale model was finished, I tested it in the wind tunnel of the Guggenheim Aerodynamic Laboratory at New York University. We carefully observed the air flow and stopped the tests at appropriate times, quickly modifying the smoke-deflector contour with malleable clay, and starting the windstream again until we obtained optimum results. This device was incorporated in the actual locomotive, built of steel sheet metal. The T-1 was also equipped with this airfoil device. The S-1 has become a classic, and it has been featured in railroad-engineering textbooks here and abroad.

I remember a day in Fort Wayne, Indiana, at the station: on a straight stretch of track without any curves for miles; I waited for the S-1 to pass through at full speed. I stood on the platform and saw it coming from the distance at 120 miles per hour. It flashed by like a steel thunderbolt, the ground shaking under me, in a blast of air that almost sucked me into its whirlwind. Approximately a million pounds of locomotive were crashing through near me. I felt shaken and overwhelmed by an unforgettable feeling of power, by a sense of pride at the sight of what I had helped to create in a quick sketch six inches wide on a scrap of paper. For the first time, perhaps, I realized that I had, after all, contributed something to a great nation that had taken me in and that I loved so deeply. And I had come a long, happy way myself from my start in fashion advertising. I had found my way of life and felt thankful to all those bright men who had transformed my sketches into reality. It is not maudlin to remember moments like this: they happen so infrequently.

Design for locomotive, 1937.

The Pennsy K4S was a success, and it led to the design of the S-1, an advanced expression of locomotive streamlining. Here I am in front of the S-1 locomotive, Wilmington, Delaware, 1937. The design became the basis of a large advertising and public-relations effort. Reproduced all over America on posters and schedules, the K4S/S-1 shape helped the Pennsylvania Railroad establish a reputation for advanced engineering and design.

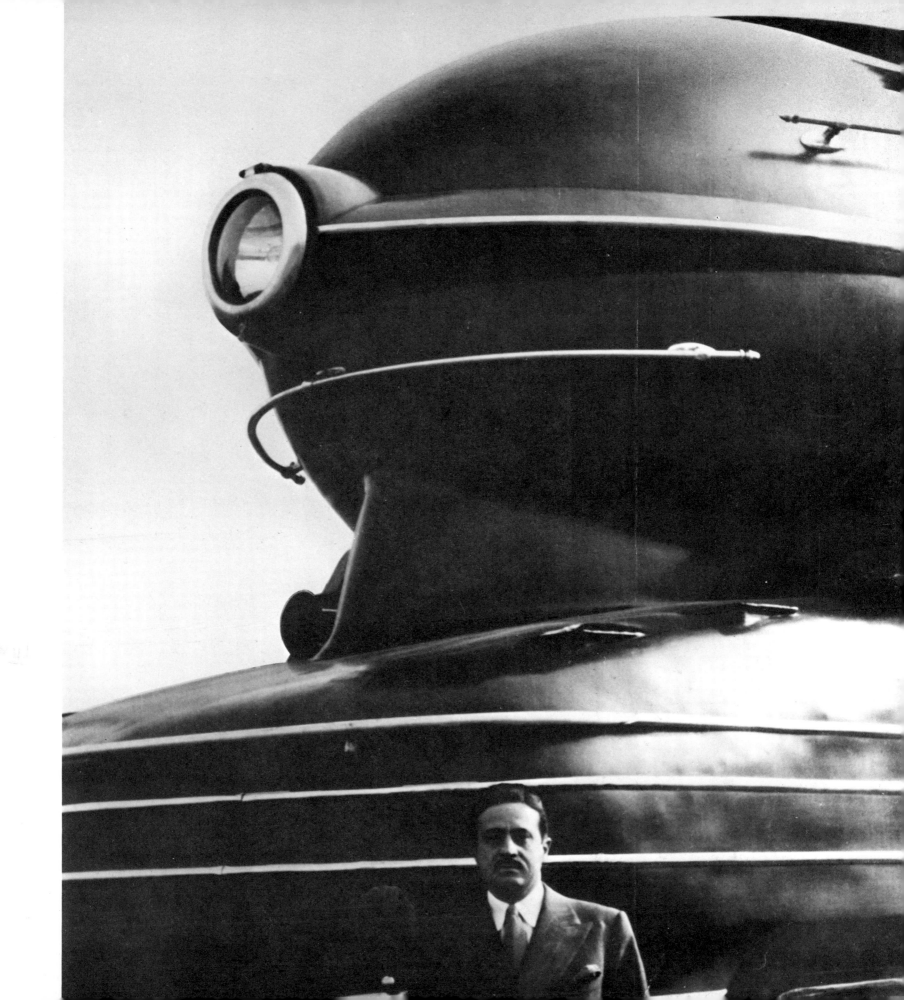

When the Virginia Ferry Corporation, operating between Norfolk and Cape Charles, needed a new vessel in 1933, I thought it would be an occasion to create something seagoing that was truly modern, and to establish new standards for this type of craft. The company was closely affiliated with the Pennsy, whose executives were beginning to appreciate what I was doing in the railroad field. So I was given the chance to work on the vessel's silhouette and its interior appointments.

The results surprised a lot of people, both in the shipbuilding and business worlds. People liked its unusual (then) appearance and some of its peculiar (then) appointments. A small band, at my request, was placed on board in the lounge; passengers boarded the ferry now as much for fun as to get from one point to another. They danced, went to the snack bar, and made the round trip as though they were taking a short pleasure cruise. The *Princess Anne,* therefore, carried, besides great numbers of cars and trucks, a lot of people having a great time; it was all very gay and romantic, a long way from a boring ferry crossing. It even turned into a huge financial success for the operators. The elongated openings on the

The *Media Luz,* 1937.

promenade deck made the ship look longer, and, when illuminated at night, the *Princess Anne* looked like a giant liner ready to cross the Atlantic.

The ship's look and additional usage led the *MIT Review* to publish photographs of it with the following caption: "FERRY FAIRED: Raymond Loewy, the industrial designer, ended up with this steamer when he set about, at the behest of the Virginia Ferry Corporation, to streamline a bay boat working out of Norfolk. Are ferries leading the way to streamlined liners?" They did lead the way. From then on, ocean liners were influenced by the *Princess Anne*'s streamlined approach. Besides better appearance, the new silhouette had valuable repercussions in the public-relations area, and large navigation companies used it in their promotional and advertising literature for all new ships. The smooth, flowing lines of the superstructure facilitated maintenance. At night, when flying dozens of colorful flags and burgees illuminated by powerful floodlights, the *Princess Anne* looked like a queen, a glorious happy ship. Colors of the hull and stack were white enamel and cobalt blue. The exhaust stack was trimmed with stainless-steel strips and the company's insignia, a large disk of blue with gold-leaf accents.

Princess Anne's success led me to apply the same design concept to larger vessels. This sketch made in 1938 was my idea of an ocean liner that anticipated by forty years the outline of some recent ships.

Princess Anne.

Carried away by the *Princess Anne* look, naval engineers in America and overseas went on a binge which produced floating monstrosities. Here are a few examples of what uninformed rank imitation can generate. I wonder sometimes whether I should have designed the *Princess Anne* and thereby become an accessory before the fact for such seagoing infractions of good taste.

Modernistic

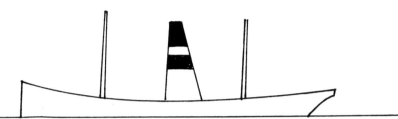

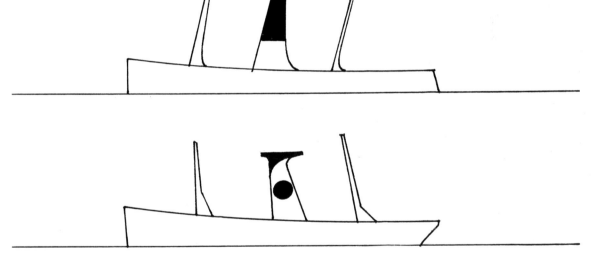

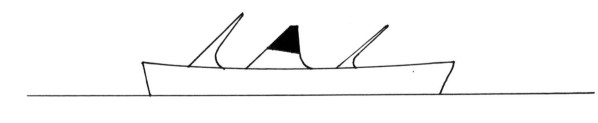

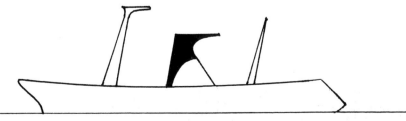

Different sketches for exhaust stacks.

COLDSPOT

The Sears Coldspot, 1934, is generally recognized as the classic case history of industrial design applied to mass production; it marks the beginning of the profession in America for three main reasons: First, the phenomenal success of the new design; yearly sales went from 65,000 to 275,000 after the new model was introduced. Second, Sears' management, sales force, and even the competition, credited the improved appearance as the main reason for the increase. Third, it marked the beginning of a new model-making technique that saved enormous time and cost and has since been universally adopted for all kinds of products.

The simpler appearance, free from superfluous ornamentation, gave the new Coldspot an attractive, convincing look. Women wished to have it in their kitchens because they saw it as an expression of their "refined taste," and it cost no more than other refrigerators. The new modeling technique was motivated by the fact that sketches and renderings necessary at the conceptual stage often lose their value when translated (blown up) into large three-dimensional mock-ups. Some of the subtleties seem to disappear, and the character of a sketch may change, sometimes for the worse. Until that time large mock-ups were made of wood; fleeting design inspirations or corrections were cumbersome and slow, interfering with design spontaneity—sculptors will particularly understand what I mean.

However, our knowledge of manufacturing limitations kept us from trying unrealistic shapes. A blending of aesthetic concern and hard-core practicality is at the heart of industrial design. Computer designing or slide-rule solutions invariably show through; they have a sterile look that one senses at a glance. What I needed was a system whereby an intuitive design idea could be tested quickly, visually, and by touch. I have often relied on "feel," sometimes with my eyes closed; sensory impression often helps to evaluate the finesse of a subtle form, or lack of it.

The answer was the modeling clay that I had already used in the Gestetner case and later in building automobile mock-ups. The clay eliminated the difficulty of imparting to a carpenter design subtleties that had to be made on the spur of pure intuition. This technique gave the design flair and spontaneity.

So we started making wood box frames ("bucks," in designers' jargon) slightly smaller than the finished refrigerator, and we covered them with thick layers of modeling clay, giving us blocks of malleable material to work on; then we could carve or sculpt any shape we wished. It became apparent that the clay was easier to work with when warm, so we developed an electric clay-warming oven. When the models were ready for presentation to the client, we naturally wished to place them under correct lighting and display them in the right sequence. We discovered that they were too heavy to be moved without damage. So we built the next batch of mock-ups on low platforms mounted on roller bearings, permitting easy movement.

The hardware, such as hinges, latch, name plates, etc., was carved in wood and sprayed with aluminum lacquer, but we felt this lacked sparkle and looked clumsy. So we sprayed on a thick layer of copper paint. This metal finish was then polished carefully and the hardware placed in plating tanks; as a result we had parts that were perfectly nickel-plated and highly polished. The final object looked and felt exactly like a solid metal piece of hardware, but it cost only a fraction as much and could be produced quickly. This was a major step forward.

When we began our design, the Coldspot unit then on the market was ugly. It was an ill-proportioned vertical shoebox "decorated" with a maze of moldings, panels, etc., perched on spindly legs high off the ground, and the latch was a pitiful piece of cheap hardware; we solved all these problems in no time. The open space under the refrigerator was incorporated into the design and became a storage compartment. The new latch was substantial as well as attractive (like the door handle of an expensive automobile, the hinges were made unobtrusive, and the name plate looked like a piece of jewelry. The new design connoted quality and simplicity.

Shelves had been a serious problem for the refrigerator industry. They had to be formed of metal wire and assembled by hand, welded and dipped in rust-proof finish. They rusted just the same, and their cost was exorbitant due to the excessive amount of labor involved. And they always looked messy. At the time, I was working in Detroit on the front-end design of the Hupmobile. Among the materials considered for the car's radiator grille were several samples of perforated aluminum. In order to give strength to these large panels, they were reinforced at the back by longitudinal ribs. The whole thing was extruded in one neat and inexpensive piece. While looking at it, it struck me as being the perfect material for my Coldspot shelves. I ordered several samples with the right perforated slots, had them cut to the dimensions of the Coldspot shelves, and took them to Sears Roebuck. They were a sensation—here was our answer; attractive, foolproof in manufacture, and completely rust-resistant. These shelves were adopted, imitated by the industry, and have been standard ever since. The discovery was, incidentally, a windfall for the aluminum industry, which supplied the metal for several million shelves each year, and it remains a classic example of design cross-pollination.

COLDSPOT

REG. U.S. PAT. OFF.

"Super Six"

Lovely Modern Design
Super-powered "Package Unit"
Full 6–cubic foot size
About half usual price

A NEW COLDSPOT for 1935 and a NEW Standard of Value in electric Refrigerators. By Value we don't mean just a lower price. You will never appreciate the Value offered in this COLDSPOT merely by looking at its price. Here is all we ask: Forget the price for the moment and consider this COLDSPOT purely in terms of Quality. Study its Beauty. Check its features. Analyze it strictly in terms of what it offers you. Then compare it with any other refrigerator of similar size, selling in the $250 to $350 class. We say that you will find the COLDSPOT actually a *Better* refrigerator, *In spite of the Fact That It Costs Only About Half as Much.*

USE YOUR CREDIT. You don't have to pay cash. See Easy Payments Prices and Terms on **page** at right.

All Prices for Mail Orders Only.

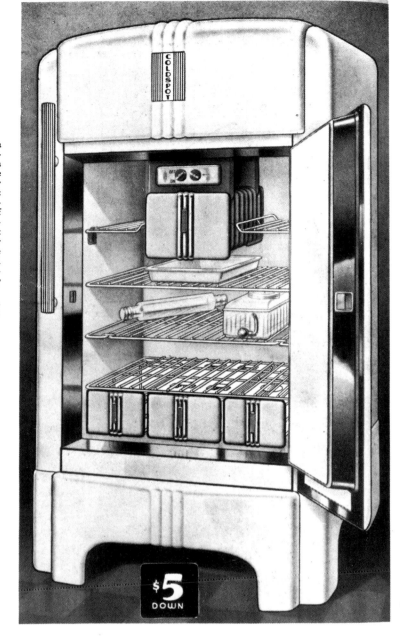

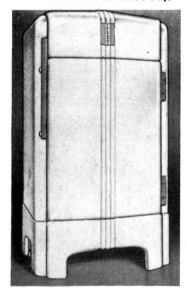

VEGETABLE FRESHENER

Large, covered, porcelain enamel vegetable freshener for keeping lettuce, celery, tomatoes, etc. in a fresh, crisp condition. Easy to keep clean and sanitary. Slides in and out exactly like a drawer.

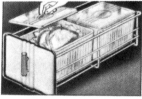

STORAGE BASKET

Large wire basket, containing two oversize covered glass dishes to keep butter, salads or left-overs from absorbing the taste of other foods in the box. These dishes can be removed for kitchen use if desired.

STORAGE BASKET

An open wire basket for holding coarse vegetables, fruits, etc. to eliminate breakage. (This container and the 2 shown at left suspend from lower shelf like drawers.)

WATER COOLER

Covered glass water cooler with down faucet. Holds about a gallo liquid. Can be used for iced tea le ade or other beverages. Especially able during the hot months.

For the first time, tough, down-to-earth Sears Roebuck went "artistic," and millions of eyebrows were raised. The Sears catalogue unblushingly printed: "Lovely modern design," "Study its beauty," "A better refrigerator, in spite of the fact that it costs only half as much." That last item was a bombshell. Discovering that improved appearance can reduce cost, industry started to look our way.

COLDSPOT

When we began our design, the Coldspot unit then on the market was ugly. It was an ill-proportioned vertical shoebox "decorated" with a maze of moldings, panels, etc., perched on spindly legs high off the ground, and the latch was a pitiful piece of cheap hardware; we solved all these problems in no time. The open space under the refrigerator was incorporated into the design and became a storage compartment. The new latch was substantial as well as attractive (like the door handle of an expensive automobile), the hinges were made unobtrusive, and the name plate looked like a piece of jewelry. The new design connoted quality and simplicity.

Shelves had been a serious problem for the refrigerator industry. They had to be formed of metal wire and assembled by hand, welded and dipped in rust-proof finish. They rusted just the same, and their cost was exorbitant due to the excessive amount of labor involved. And they always looked messy. At the time, I was working in Detroit on the front-end design of the Hupmobile. Among the materials considered for the car's radiator grille were several samples of perforated aluminum. In order to give strength to these large panels, they were reinforced at the back by longitudinal ribs. The whole thing was extruded in one neat and inexpensive piece. While looking at it, it struck me as being the perfect material for my Coldspot shelves. I ordered several samples with the right perforated slots, had them cut to the dimensions of the Coldspot shelves, and took them to Sears Roebuck. They were a sensation—here was our answer; attractive, foolproof in manufacture, and completely rust-resistant. These shelves were adopted, imitated by the industry, and have been standard ever since. The discovery was, incidentally, a windfall for the aluminum industry, which supplied the metal for several million shelves each year, and it remains a classic example of design cross-pollination.

This model created ever-new sales records and staggered the competition: appearance was now definitely proven to be a sales asset. Coldspot's prestige went sky high, positively affecting the sale of all kinds of *other* Sears household products. So much of this seems obvious now; it wasn't then.

Another improvement was the "feather touch" latch. This latch was designed so that a housewife with both hands full could still open the refrigerator by pressing slightly with her elbow on a long vertical bar. In addition, the latch was connected by remote control to a small foot-operated pedal close to the floor. All these combined features made perfect material for the advertising fellows and supplied the salesmen with great pitches.

These results could no longer be ignored. The subject became part of studies in universities and was included in the program of marketing and merchandising seminars. Large corporations studied the Coldspot story, and the lesson was not lost on sales executives everywhere.

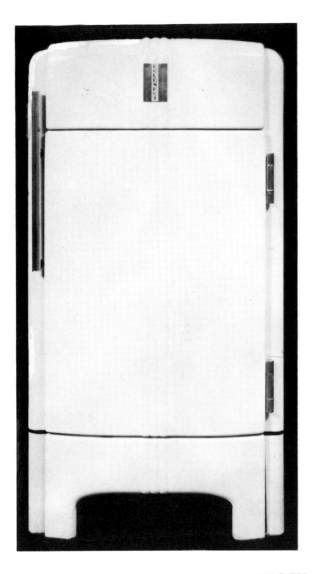

As early as 1935, I felt that industrial design could greatly improve the design, layout, and planning of department stores, while coincidentally supplying more attractive décor, better illumination, saner traffic, and more accessible stock storage. I placed William Snaith in charge of a new retail-store design division, and soon we were swamped with assignments. Such men as Minoru Yamasaki, Gordon Bunshaft (who later became president of Skidmore, Owens and Merrill), and dozens of others who became highly successful, started out quite simply as draftsmen in our organization. Over the next ten years we became recognized for our innovations and were retained by department-store managements as far afield as France, England, Holland, Belgium, Germany, Italy, Switzerland, and Mexico.

A project of ours in which I took a large personal interest was the new Saks Fifth Avenue store. Horace Saks wanted to make it New York's most fashionable department store. As a move in that direction, I suggested that the rather drab elevator operators be replaced by handsome young men wearing black tuxedos, white gloves, and white Ascot ties. The effect suggested Saks' new sophistication, but this was only symbolic of wide-spread functional proposals we made to Saks and which they instituted.

To describe our influence upon the very *concept* of department stores and shopping centers alone would require a volume on fashion, design, and merchandising. The following list of clients will give an idea of the role we played in this field: Abraham & Straus; Au Printemps N.A.; Bloomingdale's; Bon Marché; Coop; Dayton Company; Famous-Barr; Galleries Lafayette; Grand Bazar de Liége; Hahne & Company; D. H. Holmes Co., Ltd.; Horton; J. L. Hudson Company; Hudsons Bay Company; Innovation; John Lewis & Company; Lord & Taylor; R. H. Macy & Co.; Joseph Magnin Co.; Neu Warenhaus A.G.; J. C. Penney Company, Inc.; Priba; Prisunic; Saks Fifth Avenue; Stix-Baer & Fuller.

The Lord & Taylor branch store in Manhasset was the first American experiment in large-scale, suburban department-store merchandising, and it created a new trend in America and abroad. In most cases, we designed the structure itself through renderings and scale models, then turned these over to registered architects on our staff who took over, "engineered" the project, and filed the plans. Our professional relationship was effective and pleasant. William Pereira is one example among many of the leading architecture firms we used, as is Walden Beckett, both of California.

Foley's—Houston, Texas. We soon found out that efficiency demanded that back-up merchandise be adjacent to each department. The advent of air conditioning made it possible to eliminate windows and to use the enlarged wall surface as storage space. It also gave us the opportunity to improve the quality of lighting at a constant level, independent of the time of day or weather conditions—a cheerful ambiance could therefore be present throughout the day.

Lord & Taylor—Manhasset, New York.

Woodward & Lothrop—Washington, D.C.

We were retained by the Panama Line as designers and placed in charge of all interior décor, lighting, and furnishings. Three ships were engineered by George Sharp, the foremost American naval architect, with whom we worked closely. Experts in naval engineering recognize that the interior treatment of these ships created a new orientation; for the first time American vessels displayed a contemporary décor free from Louis XVI, Old English, or baroque influence. A new naval décor—reflecting the modern world, efficient, comfortable, and cheerful—replaced the often old-fashioned drab and stuffy pseudo-luxury of bygone eras. Luxury and modernity were not at odds. The reaction was immediate and recalled the influence of the *Princess Anne* ferryboat upon the smokestack design of modern vessels. The S.S. *Panama* later was requisitioned by the government and became General Eisenhower's headquarters during the invasion of Normandy. Even earlier, my colleagues Walter Teague, Henry Dreyfuss, and I were involved in the interior layout of General Marshall's inter-allied command headquarters in Washington, and we also designed much of the equipment. I regret that lack of photographic records makes it difficult to display here the contemporary ambiance we created for American shipping.

The dining room and the bars, the staterooms equipped with built-in furniture containing drawers that glided smoothly and silently, all were appreciated by travelers during long passages.

Lighting and a sunny quality, the atmosphere colorful and gay, the uniforms of the crew well tailored. We designed all printed matter—menus, maps, etc. Each time a ship returned to its New York berth, one of our specially assigned designers would spend two or three days on board to see that everything we had designed remained shipshape. It was part of our services for more than ten years.

My association with Studebaker started in 1938 and lasted until 1962.

The keynote of my work was simplification, and these four photographs illustrate the integration of the rear lights into the increasingly effective new fenders, the absorption of the trunk into the body, simpler wheels, and more protective bumpers.

Rear-end streamline was improved, less drag.

I suggested to the Chrysler Corporation that it display an animated, realistic model of a rocket-launching spacecraft installation at the 1939 New York World's Fair as an expression of the corporation's long-range vision and technological leadership. It became one of the fair's greatest attractions. Every twenty minutes a batch of visitors was admitted to a huge half-cylindrical area, in total darkness except for the spotlighted rocket on its launch pad installed in a shallow pit on one side of the space. The launch area showed all kinds of lights, white and colored, blinking and conveying a feeling of activity and suspense. A deep rhythmic sound, the whirling of powerful motors, compressors, and high-frequency sound waves made the whole area vibrate and pulse.

The rocket seemed to be made ready for lift-off and the audience was on edge. Then at the sound of sirens: *Lift-off!* In a moment, a blinding flash of hundreds of strobe lights and the roar of compressed air suddenly released—the rocket, through optical illusion, seemed to disappear overhead in the blackness of space. Then total silence while a gradually diminishing point of light faded away to the stars.

People who saw it once often returned; it was a thrilling preview of what many believed could happen in the future. It did: only thirty years later.

At the end, a message was broadcast through a loudspeaker, stating that, in the designer's opinion, transcontinental and transoceanic mail and small valuable objects might be carried and parachuted, via recuperable missiles, to faraway cities in a matter of minutes. It still might happen.

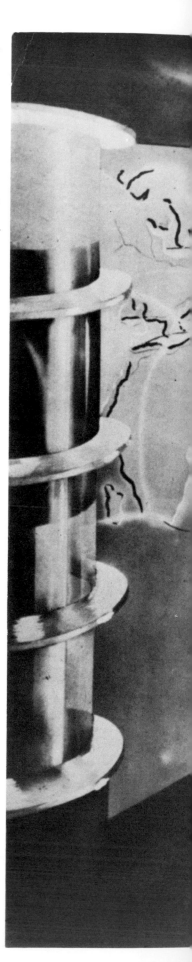

I designed this taxicab of the future in 1938 as part of the focal exhibit on transportation in the Chrysler Motors Building at the New York World's Fair. Tomorrow's taxicab will probably be a three-wheeled vehicle—one that is easy to handle. It is designed for short traffic hauls, is light, and can be readily manipulated in congested areas. In this model, I make use of curved glass for greater visibility; a sliding door simplifies entrance and exit.

It is powered by electricity—no harmful exhaust, no noise.

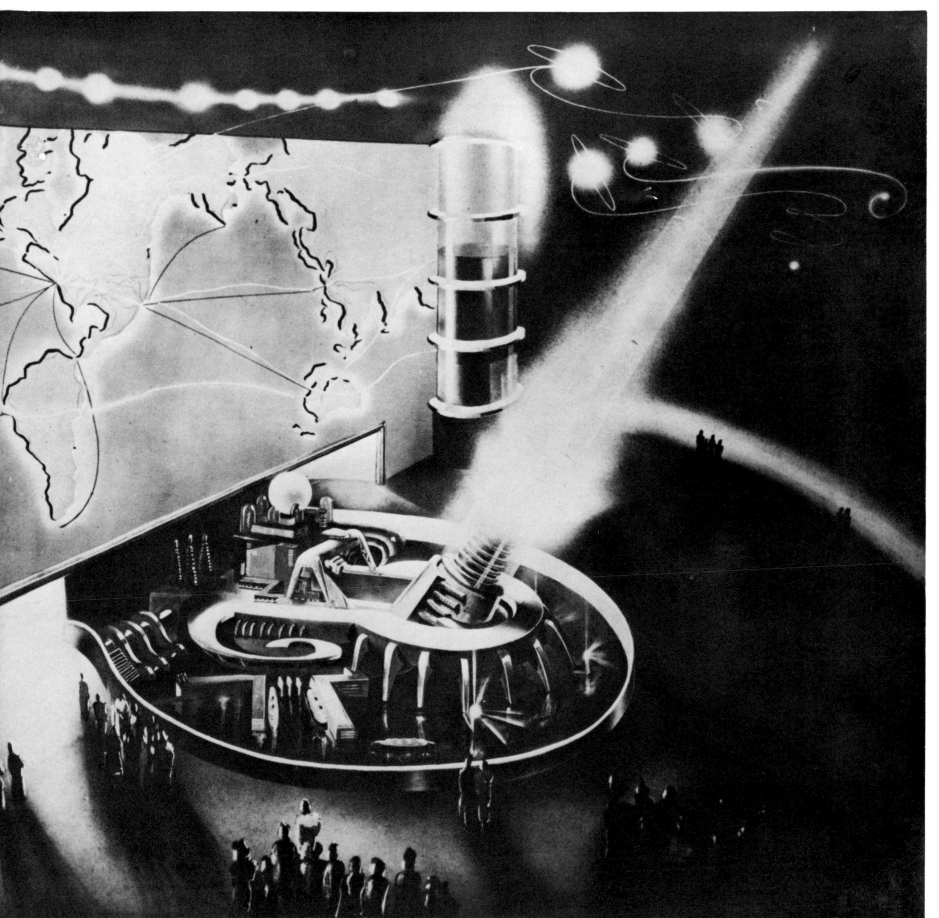

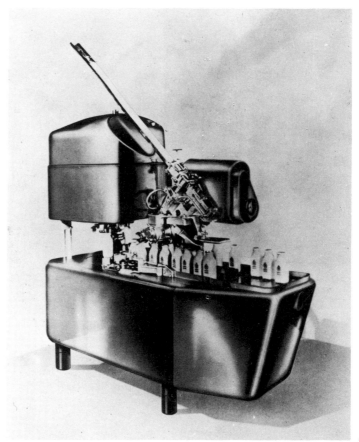

A bottle capper before and after.

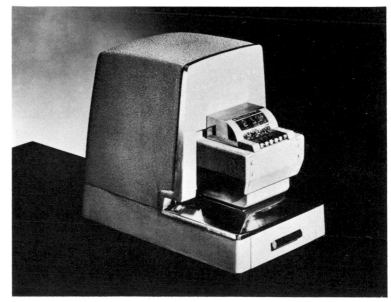

A perforating machine before and after.

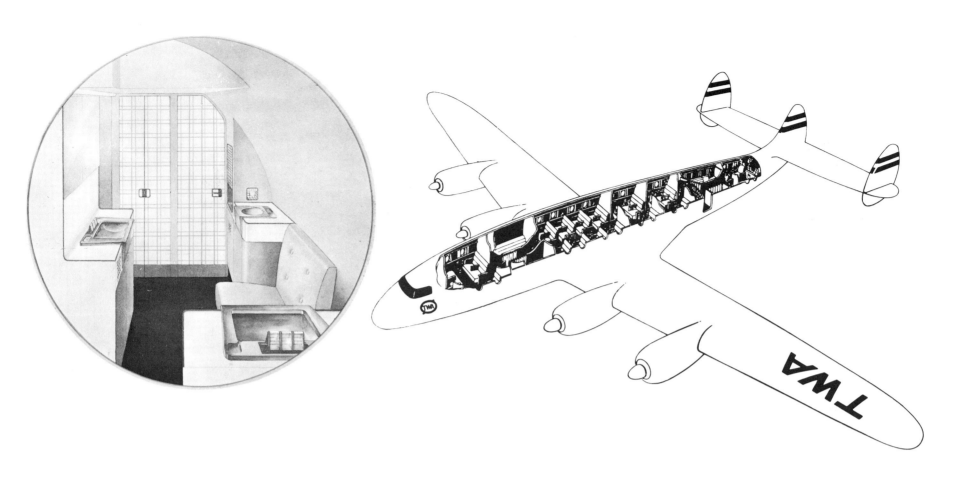

A typical sketch of the interior appointments of the Lockheed Constellation, 1945. I was retained by the president of Lockheed Aviation to design the interior of the Constellation, one of the most successful aircraft in aviation history. It had a most graceful fuselage. We made the interior cheerful and colorful to relax and reassure the many passengers who at that time were afraid of flying.

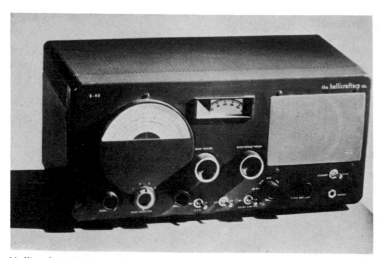

Hallicrafter, 1947.

These radios and television sets were technologically remarkable. Instead of enclosing them in gaudy ornamented walnut cabinets, with gold-fabric speaker grilles, we tried to express their qualities in a direct manner. We used black and white: dials were precise, and they looked mechanically convincing. These sets were commercially successful, an indication that a considerable number of consumers appreciated conservative design. In the late sixties, the Japanese discovered and built upon the Hallicrafter "look"; they have used it ever since. The Japanese "lead" became the international style, and this trend continues. ▶

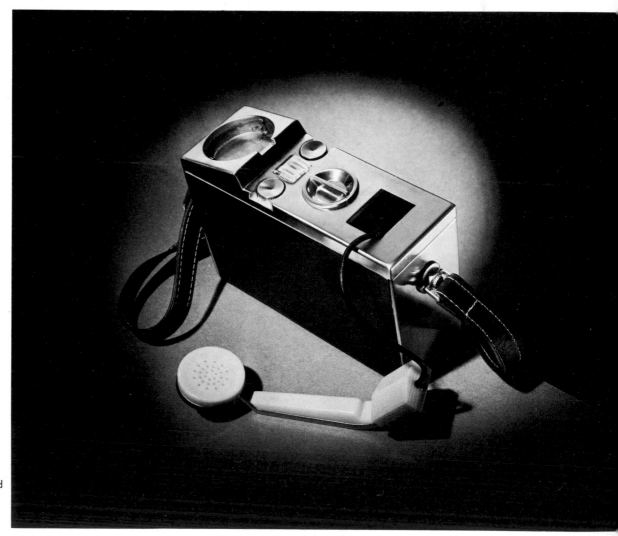

This walkie-talkie attracted much attention in the electronic industry in 1947 because its new technological qualities were effectively expressed in its straightforward appearance. The absence of "styling" was an honest and appropriate design approach for the firm, Hallicrafter.

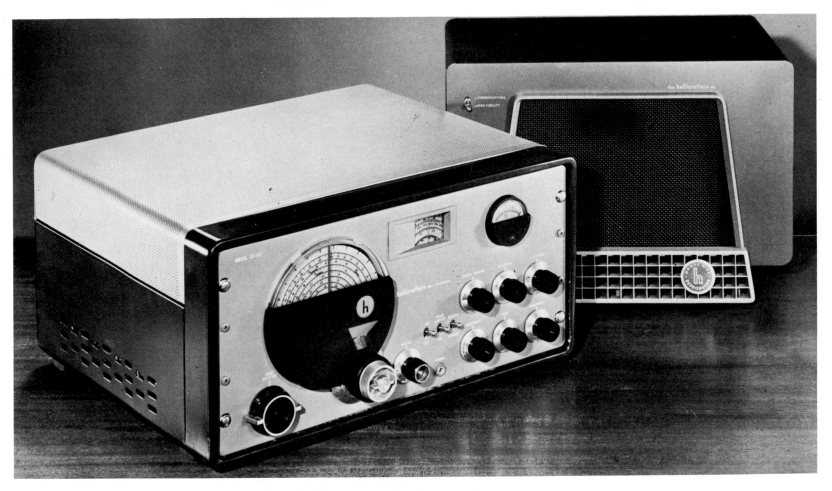

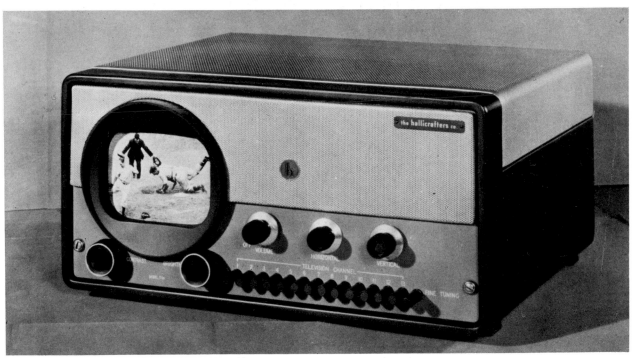

For twenty-five years I virtually commuted between New York and Dayton, Ohio, designing many different household appliances for Frigidaire. During these years sales of Frigidaire refrigerators soared.

Simplicity was the guiding rule; these units remain essentially up-to-date today because their design was basic and free from trendiness. The quaint look of the ladies in these photos suggests a change in advertising as well—they are a far cry from contemporary television and print *hausfraus* who wear almost transparent nighties while fixing breakfast . . . all to sell a refrigerator.

Neat and simple, easy to clean, with a large lower storage area, this Frigidaire hugs the floor, eliminating the problem of cleaning beneath it. Notice the slender, comfortable handle. We discovered that consumers appreciated design finesse.

A room air conditioner.

"A charming housewife gracefully displaying her Raymond Loewy design to a lady truly astonished by its irresistible beauty." —A typical copy line of the period.

Thirty years old and well-preserved for its age. The crown trademark in gold-finish metal inexpensively added, salesmen noted, "a note of distinction."

A freezer.

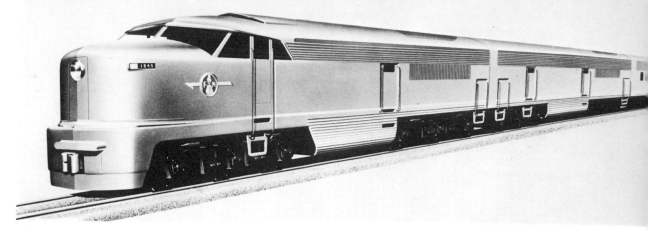

Fairbanks-Morse diesel electric, experimental, 1945.

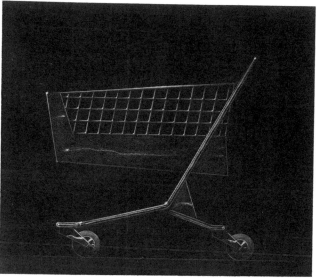

Lucky Stores shopping basket, 1945.

1942.

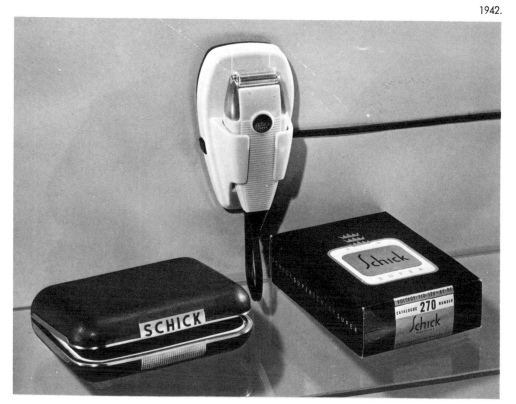

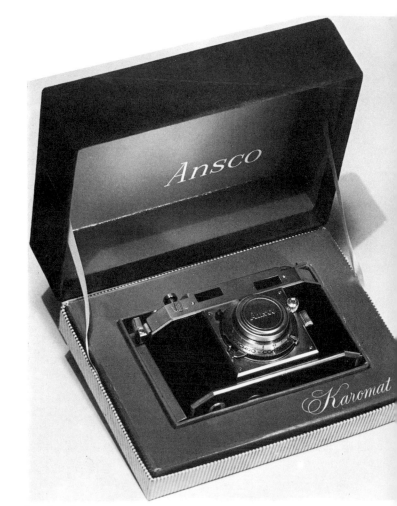

Two examples of packaging concepts.

1950.

The previous Mobil model and the new model, 1950, whose sales success indicated that attractive design improved sales of even standard items.

I was tired of radios made of sculptured walnut, with their curlicues and gold-woven cloth speaker covers, so I built a radio to please myself. Its components were displayed within a box of clear plastic on a black base. Control knobs, switches, and dials were designed honestly, the speaker grille was satin-finish perforated aluminum, and the whole set had an appropriate electronic look. The pivoting directional antenna was mounted on a vertical jack inserted in an outlet, and the antenna was removable for shipment. These design elements contrasted strongly with the product look of the period.

Lucky stores supermarket, 1945.

Cream separators were hard to clean, due to their many recesses, cavities, and exposed areas. I believed impeccable cleanliness was essential in such a machine and designed the improvements for McCormick-Deering in 1945.

The International Harvester Farmall tractor before it was redesigned. Mud-clogged wheels became heavy and hard to clean. A fair number of farmers found it difficult to reach the high seat. Whoever designed it must have forgotten that some farmers are old, or paunchy, or arthritic. Industrial design means a concern for the broadest spectrum of humanity, not only for ideal forms. Farmall was also rather unstable, due to its three-wheel arrangement.

The new Farmall, 1940, with an easier-to-reach seat, easier-to-clean wheels, and increased stability based on a fourth-wheel arrangement. Many of the improvements were of course of an engineering nature and, therefore, credit is hereby given to those International Harvester gentlemen with whom we worked.

The International Harvester Caterpillar tractor before it was redesigned.

The McCormick-Deering International Harvester tractor, known as the Caterpillar. In its original version it was rather uncomfortable to operate, the exhausts were so short that the driver inhaled the fumes. We therefore raised the pipes, while creating a more comfortable seating and operating pattern. The sides were greatly simplified and smoothed out, which improved maintenance.

The new design, 1942.

In view of the prestige and power of International Harvester, I thought that their trademark was frail and amateurish. The firm's executives asked me to show them what I had in mind. I left Chicago for New York on the train and sketched a design on the dining-car menu, and before we passed through Fort Wayne, International Harvester had a new trademark. It was reminiscent of the front-end of a tractor and its operator. The spur-of-the-moment creation of this trademark and its subsequent longevity contradict the notion of other designers that designing new trademarks always demands thorough, lengthy, expensive research and a great many interviews, tests, and polls, plus market research and campaigns to build consumer awareness, etc., etc. These things may sometimes be necessary, but it all must start with an inspired, spontaneous idea; I still believe in intuitive trademark design, and I designed the Exxon logo decades later rather intuitively as well. Briefly, the International Harvester trademark was based on a reduction to essentials and a respect for function.

International Harvester's machines and equipment were so complex and expensive that they required tens of thousands of readily available replacement parts. They had to be well packed in cartons, bags, and containers which were both inexpensive and quickly identifiable in terms of their contents. We solved the problem after making a thorough analytical study and establishing a serialization system together with a visual identification code. This is a case in which research was paramount.

Our work for International Harvester illustrates that the application of sound industrial design techniques can work to the advantage of corporations that are already large and generally well organized. In the space of a few years, we suggested design improvements for products, service centers, showrooms, spare-part packaging, printed promotional materials and, above all, we created a trademark that is known internationally.

The prototype for the International Harvester Servicenter was designed so as to be fully standardized (modular) and easily expanded or contracted according to commercial-site requirements. The sales floor was well lighted, access to large equipment was easy, the floor simple to keep spotless. The goal was to create an atmosphere conveying a sense of product quality, longevity, and sound engineering—an honest, convincing approach in accord with the company's prestige and reputation. Over 1,800 units were built.

Singer vacuum cleaners were already known for their reliability before we were asked to design a new model in 1946. Unfortunately, however, the old model presented the housewife with problems in carrying and storing it; it also did not clean under sofas or low furniture because its casing was too high. We made it easy to transport; it could be hung in a closet; and the cleaning head was no higher than a cigarette pack, so it could clean under practically anything in the home.

The exploded view below shows how easy it became to service or maintain the vacuum cleaner. An intense light source incorporated in the leading edge, like the headlights in a car, made cleaning in dark places even easier. As to the Singer trademark, it was, as was the Coldspot's, well located and rather deluxe, a suggestion of the product's quality.

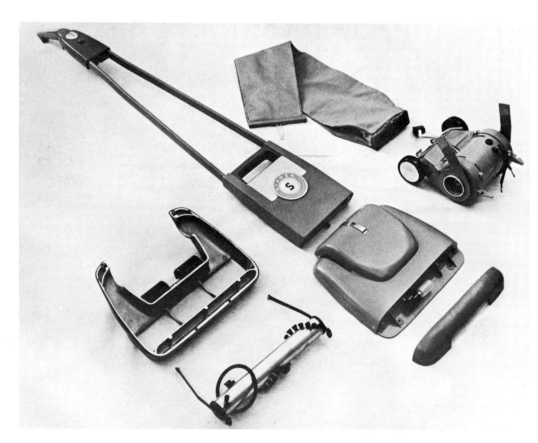

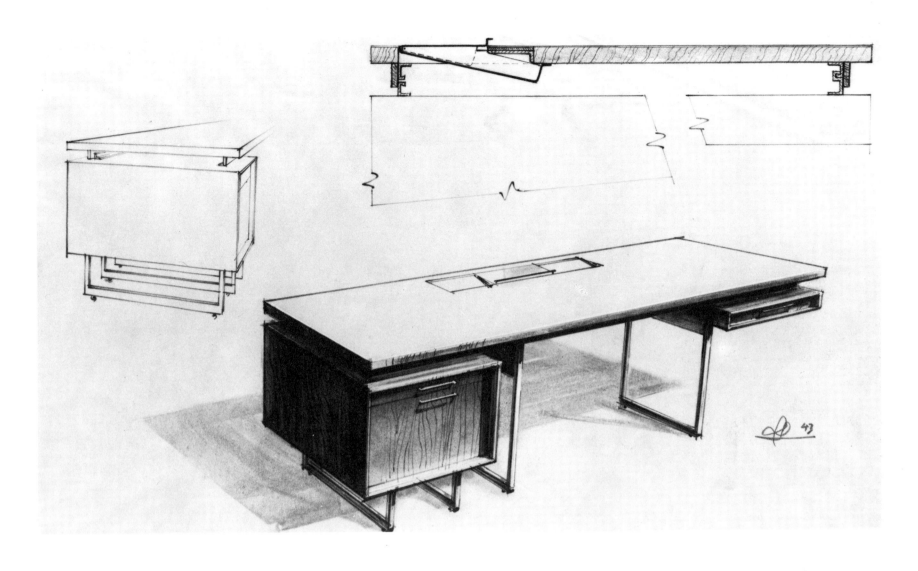

These 1943 desk designs were a sharp departure from the usual flat oak or roll-top models. Our task was to make more functional, spotproof, and scratch-proof desks, equipped with silent, easy gliding drawers. In those days some secretaries preferred desks with baffle-type panels on one side to preserve their modesty; recently I have noticed the disappearance of these panels in a more liberal time.

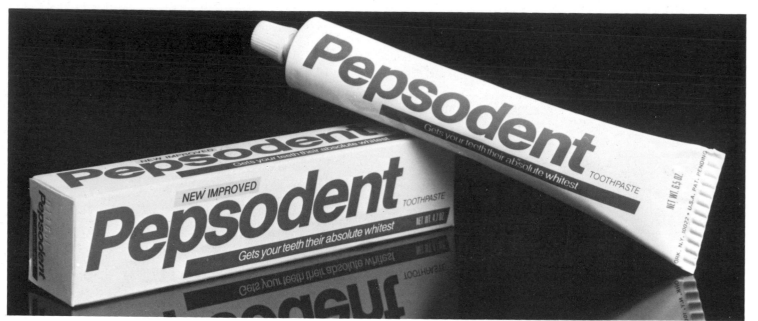

Product packaging.
1945.

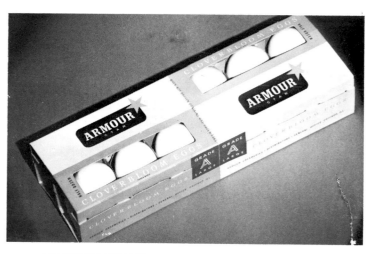

1944–1945.

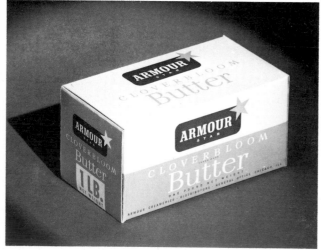

For the larger vehicle-design task, I rented
a large, high-ceilinged, empty store at the
corner of Park Avenue and 45th Street, a
former power-boat showroom. There we
installed the proper machine tools and built
a full-size mock-up complete with seats,
washroom, etc. It was the first Greyhound
double-decker—later named the Scenicruiser.
We reinforced the lower part of the body at
collision level as a safety measure. Also for
the purposes of safety, I placed a large white
disk with a bright red arrow on the inside of
the door; when the door opened, the arrow
pointed down to the steps. As seat upholstery, Greyhound asked us to use a certain
type of fabric with a pattern that disguised
spots. After Greyhound let us know which
certain colors were most common, we designed
a small pattern with irregular shapes in
matching shades, as a kind of disguise. A
color rendering of the 1946 Greyhound bus
is on page 65.

Thanks to Orville S. Caesar, Greyhound's
chief executive, an imaginative and trusting
man, the modern Greyhound bus was developed. Our plan was to build it on a GM
chassis. My first Greyhound assignment was
given to me in 1933, when I met Caesar and
mentioned that the silhouette of the Greyhound shown on buses of the period suggested a fat mongrel; amazingly, he agreed
with me and asked me to do better. I got in
touch with the American Kennel Club, which
sent me a picture of a thoroughbred greyhound; silhouetted, it developed into today's
logo.

In 1943, I bought a Cadillac convertible and rebuilt its body at a New York coach builder. I replaced the heavy grillework with a light aluminum perforated panel. On the hood I placed a thin chrome-plated circle from a Frigidaire with a gold-plated escutcheon in the center. Another ring was placed on the door. More important, of course, was the entirely new fender line that visually seemed to lengthen the body and provide a feeling of speed. Five small, simple louvers were added. The body color was light tan, my favorite car color, and the upholstery was the same shade. Three years later, Buick brought out a model that virtually duplicated this fender-line concept, and it made a big hit with the customers. Although there are fortuitous design coincidences, the styling was so similar as to have made me wish I had taken out a design patent!

Although I did not design the Coca-Cola logo, I redesigned the red dispensers installed on counter tops in soda fountains, cafeterias, drugstores, snack bars, etc.

New design for delivery truck, with slanted shelves on both sides, facilitating handling.

Coca-Cola also asked us to redesign its bottle to provide a slenderized look. We also designed the large red coolers used in stores and cafeterias as well as new cans and labels, six-packs, cartons, etc. An interesting problem was the design of a prototype delivery truck, so that heavy cases of cans or bottles could be handled quickly, easily, and safely. Another task was to design a quality-control unit to be used in Coca-Cola's bottling plants. Coca-Cola remained our client for over ten years.

I was often frustrated by working for Detroit, and sometimes by simply observing its designs. Sketching ideas for experimental cars gave me pleasure and released pent-up pressure. Forty-seven years ago, the down-slanted hood, the integrated bumper, and the replacement of the grille by an air scoop (near the ground where it is more effective) were all new ideas. A common device on racing cars, the lowered air intake ventilated the engine compartment most efficiently. The streamlined front and rear ends and the slanted (wedgelike) superstructure were based on aerodynamic principles and adapted for all contemporary automobiles. A ventilator inlet above the roof at the rear, far from the ground, injected clean, fresh air into the body instead of sucking in exhaust carbon monoxide from the car in front. The body's pillars were designed for maximum resistance in case of roll-over collisions. There were three seats forward, as in the recent (1976) French sports car, Matra's Baghera. Wheels were the chrome-disc type, and a stainless-steel strip ran from fender to fender as a protective surface against dents and scratches. The windshield afforded better visibility, and the upper part, made of tinted, heat-insulating glass, left overhead traffic lights easily visible while reducing the glare from approaching headlights. Small headlights equipped with high-intensity light sources, similar to those used in projection equipment, supplied intense narrow-beam illumination. Note the extra-wide tires now used on most sports cars.

Public-relations copy written at the time:

A CONCEPTION OF AN AUTOMOBILE OF THE FUTURE . . .
designed by Raymond Loewy Associates
1942

This design, indicative of postwar developments being considered by the designer for the automobile industry, represents a step in the direction of clean, uncluttered body styling. Visibility is increased; the power to weight ratio is improved by specifying that the automobile be constructed of light-weight metal alloys and by other technical means. Flexible seat mountings are to be desired; and the seating arrangement, three in front and two in the rear, is an innovation, permitting a narrower rear end to reduce drag. The interior can be pressure-ventilated, and the new lens-type headlights reduce highway glare and increase the range and concentration of road lighting. Projecting hardware is eliminated wherever feasible, and a faired-in undercarriage reduces drag and noise caused by air friction.

This semiconvertible, shown with the rear section halfway down, had features somewhat similar to the design above. The perforated pillars, inspired by aircraft technology, were lighter and made of highly resistant stainless steel, as was the thinly corrugated roof. Headlights were concealed, and the ovoid, conical element between the two forward bumper units was also made of bumper stock. Both cars were equipped with flush door handles to reduce projection and therefore lessen the chance of accidents to pedestrians. Notice that the center hood panel is continued inside the body and becomes the instrument panel, equipped with extra-large, easily readable dials with thick pointers.

The influence of Studebaker styling upon the automotive industry continues to be analyzed and written about.

Esquire named Studebaker as one of the dozen cars having the highest resale value and, indeed, vintage Studebakers have become collectors' items. The famous Le Mans race driver John Cooper Fitch wrote in *Esquire* in 1960:

The 1947 Studebaker bore the clear imprint of Raymond Loewy's designing genius. Its lines were graceful, long and a little hungry looking, compared with the bulging smugness of prewar cars. There are those who prefer Loewy's '53 model, but, relating each to its time, I must stay with the '47 for that was the first quivering step towards the future and an opening dividend on that promised world-of-tomorrow.

Much credit for this must actually be given to Paul Hoffman, then Studebaker's president, a remarkable man who entrusted me with the styling, as well as to the gifted design team that worked with me. For the first time, America built a truly sporty car; it made a hit, and Studebaker acquired the youth market which had previously eluded the firm's marketing efforts. My decades with the company were exhilarating and unforgettable, and my respect for its engineering department immense. I leave it to others to uncover the reasons why such a a great, prestigious firm, having at last found its market, finally disappeared at a time when it was admired throughout the world and when the Avanti had just come out with a backlog of orders. It was an industrial tragedy.

Studebaker Champion, 1942.

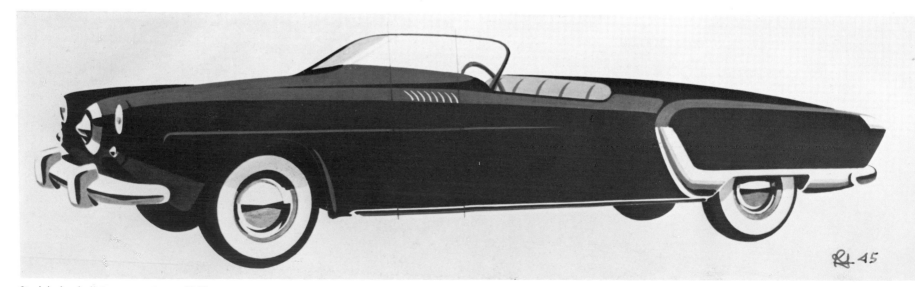

Studebaker bullet-nose sports car, 1945.

transformed Studebaker with lucite roof, c. 1945.

After I designed the interior of the S.S. *Lurline* for the Matson Line, I sailed on her maiden voyage to Honolulu. Below is a U.S. landing vessel used during the world war. I redesigned its exterior and designed all of its interior appointments. Renamed the *Carib Queen,* it became a very popular cruise ship. Even naval architects could not recognize the old warship in her new dress.

An experimental drawing for a large, fast trans-Atlantic liner that could, if necessary, be transformed into an aircraft carrier. Shown flush on the flat landing deck, two elevators are outlined. Just in front of one of the exhaust funnels, provisions were made to raise an elevator-operated control tower. A dream perhaps, but nevertheless interesting.

A typical *Lurline* stateroom with many design connections to what one finds on cruise ships today.

I rebuilt two Lincoln Continentals from new designs at a coach builder in Philadelphia, one for my own use as a chauffeur-driven town car (the original sketch is at the bottom), the other for a friend.

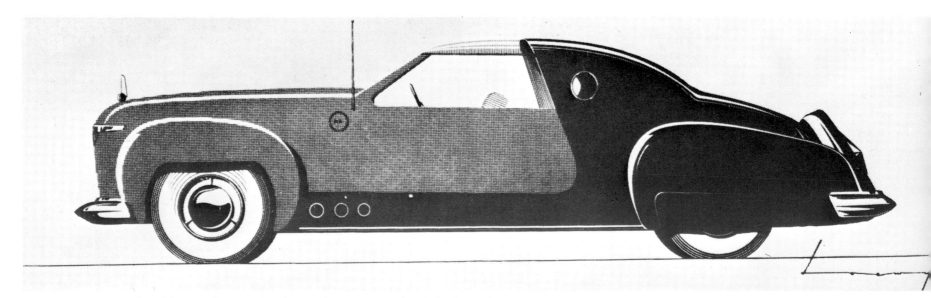

Design departures included the smooth rear section showing the integration of trunk, fenders, and rear pavilion into one flowing shape. The porthole was blue glass, the hood ornament and roof section over the driver's seat were clear lucite.

The car had plain, chrome-plated disc wheels. The monogram "RL" on the hood was in the center of a thin Frigidaire gold-plated circle. The interior was pale beige broadcloth with a black velvet, deep-pile rug.

A friend of mine in Long Island bought the second car. Now powered with a modern 8-cylinder motor, it is still often seen in New York traffic. *Ford Magazine* published its picture years later as one of the best Lincoln Continental transformations.

Bronze paperweight of advanced sports-car design presented to President Truman.

I was retained by Austin of England to develop a new prototype. My London office offered several scale models. This mock-up, seen thirty years later, is still interesting, due to its sloped fender line and hood, plain disc wheels, and slender superstructure. Once again, note the simple grille, incorporated headlights, etc. Three years later Austin had still not made the full-size mock-ups and we felt frustrated. No one in the engineering department could provide an explanation, so I went to see the firm's head. With great charm, he told me that we had been retained to expose Austin to the newest designs Americans could produce. That had been the *entire* purpose of our work. Austin, he said, however, wanted what he called a traditional "English look," and his designers and engineers now knew what the American competition would offer in coming years . . . and could counter it!

I said I understood and resigned on the spot.

We were immediately retained by Sir William Rootes as designers for Humber, Hillman, Minx, and small trucks. We also designed special Rolls-Royces; Thrupp and Maberly built the bodies. ▶

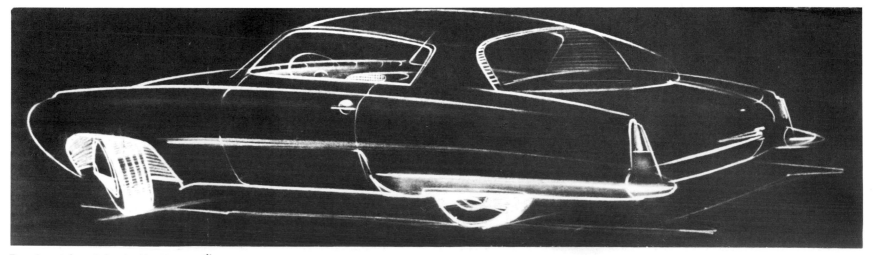

Experimental car I sketched in this period.

Rear view of one of our suggestions to Austin.

Flight bath scale for Borg, 1952.

Combination bottle-cap opener with ice-cracking sphere, 1952.

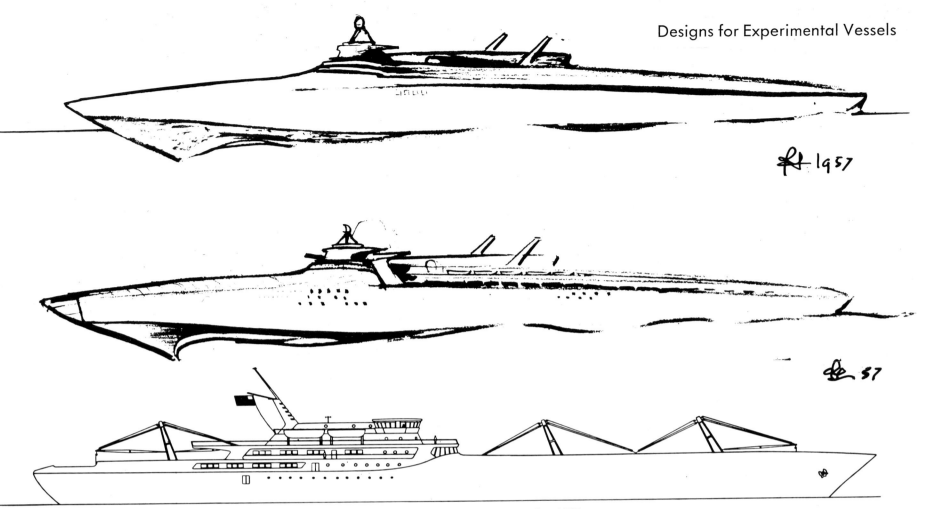

Studies for a nuclear-powered combination passenger-cargo ship, an assignment from the U.S. Maritime Administration, 1957.

Hydrofoil, 1952.

Flying catamaran outrigger racing boat, 1952.

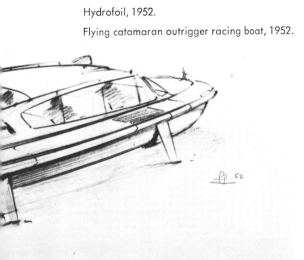

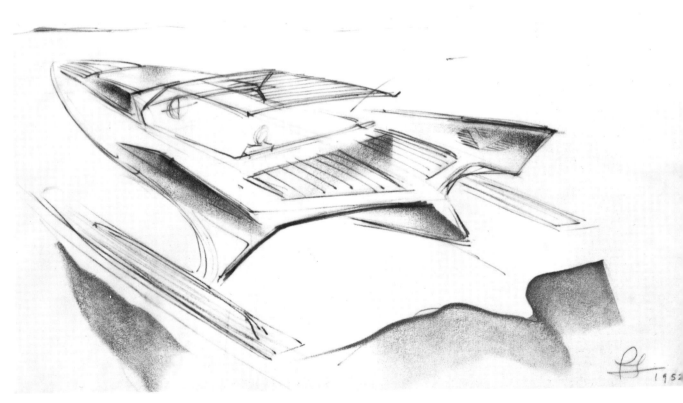

WEIGHT IS THE ENEMY

Thanks to Paul Hoffman, I was given the opportunity to design cars liberated from most of Detroit's atavistic influence. No more inbred, incestuous designs; instead, a fresh, new approach for a century-old respectable firm was demanded. The body-styling division which I formed at the plant and that bore my name became known in the profession for its talent, spirit, and sense of mission. Morale was high, the ambiance pleasant, and the designers, workmen, and modelers were arguably the industry's best. Seven or eight months each year I commuted by train from New York, about once every ten days. I was, therefore, well-known indeed to the dining-car waiters, the sleeping-car porters, and the conductors, one of whom once said to me, "You are so often on this train, Mr. Loewy, why don't you rent a bedroom by the month?"

The task, then, that Paul Hoffman gave me was to build a car for the younger segment of automobile users. Why not a Studebaker sports car? Paul, my friend, agreed to it. He immediately informed the engineering department, and our design team started on the assignment. Instead of reporting to committees and sales managers, as captive designers were obliged to do at GM, Ford, and Chrysler, we reported to no one. Bright young engineers, including Churchill and Hardig, were as enthusiastic as we were, and we soon received a chassis blueprint from them to work upon. I selected Bob Bourke, a talented designer and a good organizer, as my assistant.

As in most successful design case histories, the work went fast and effectively; I knew what we wanted before we even started, and the entire team was supportive. The result was the Starliner, the "first American sports car," described in *Fortune* magazine as "one of the hundred best designs of modern times." Starliner had on committee look; how could it? The design direction was a personal vision.

Detroit gulped, caught its breath, and then started to copy tiny South Bend. From a *New York Daily News* editorial:

THE NEWS CONGRATULATES: *The brashest, cleverest, most effective advertisement we've ever seen in quite a while was splashed last week in various newspapers by the Studebaker Corporation. The independent automobile company roared: "Congratulations, General Motors!" in letters an inch tall. Then it congratulated GM on having designed for the future a string of cars which "closely approximate the exciting styles Raymond Loewy has developed for Studebaker." Whoever thought that ad up for whatever agency deserves a substantial raise in pay and we hope he gets it.*

I had felt for the last forty years that the American automobile was too bulky and heavy, necessitating large power plants. I lectured about it, wrote articles about it, said it on the radio, and achieved only one thing: Detroit's resentment and hostility. So when I started the Raymond Loewy Design Division at Studebaker, I ordered thirty large posters, 8 x 24 inches, printed in black-and-white block letters. I had them glued to the walls and partitions, to columns, and even to the floor (protected by a coat of transparent plastic spray). The great majority of Detroit engineers agreed with me, although management did not. Finally, when gasoline shortages developed as a consequence of the international oil crisis, Detroit was forced to accept the fact that weight *is* the enemy. In 1975, *The Washington Post* ran a two-page article about my work and the caption read: NOW DETROIT NEEDS LOEWY. The headline in *Automotive News,* the industry bible, read: END OF THE DETROIT/LOEWY FEUD.

As far as I am concerned, it never was a feud; Detroit simply had no use for my ideas as long as the public wanted large cars with all their accessories, which Detroit could sell, inducing higher price tags. And businessmen rarely express appreciation for ideas that *may* lower their income. In point of fact, they wouldn't have; while Detroit was building gas guzzlers, the Germans and Japanese, with their more modest cars, ran away with the world market!

Various stages in the development of the postwar Studebaker in my styling division:

1. A ¼-size wood frame (buck) is constructed, then covered with clay and sculpted according to a rendering.

2. The cast is painted.

3. A full-size buck is made.

4. When covered with soft clay, the buck is modeled.

5. A plaster cast is taken.

6. Several scale models are displayed.

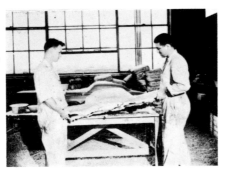

Full-size clay mock-up showing two different sides. Notice an important design feature of this car. Up to this time, body "character lines" were raised panels that gave the car a heavy look. For the first time, according to our knowledge, we adopted an intaglio (depressed) surface. Its contribution was to give Starliner its hungry and slender look.

My first Starliner at our Palm Springs home. For some unknown reason, it arrived with fancy hub caps. They were quickly replaced with the standard ones.

A white oval plate about eight inches wide must be displayed at the rear of a car's body when driving abroad. Each country has its own code letters marked in black.

The heavy black USA on top was considered by some to be rather brash and Yankee. Whether true, or important, we used less assertive lettering on all our cars simply because it looked better.

On a Studebaker chassis, I had this chauffeur-driven town car built for use in New York. Its wiring was defective and it frequently stalled on Park, Madison, and Fifth Avenues; I would get out and help our elegant chauffeur push our elegant town car out of the inelegant traffic.

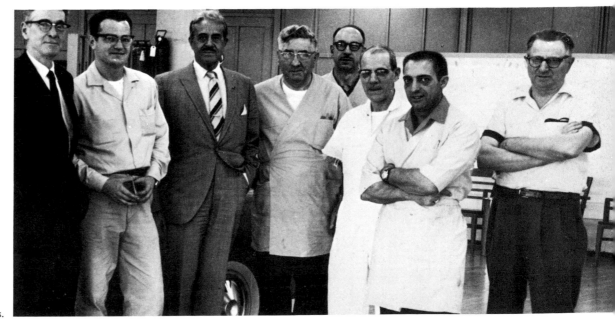

Gene Hardig, the assistant chief engineer, and the group who completed the clay mock-up. The assignment was carried out with the precision of a command operation. Most helpful during the entire project were engineers Michael de Blementhal and Frank Nemeth.

Hardig was largely responsible for management's final agreement with us that Avanti should have disc brakes, a stiff European-type suspension, and positive steering. I wished to retain the great "feel of the road." American cars were certainly comfortable then; but at the same time this comfort provided a "mushy" sensation. I felt I was sitting on a feather pillow, sliding down a mirror covered with mayonnaise. Our goal was comfort and firmness.

Testing the first production Starliner on the Studebaker testing grounds.

"The 1953 Studebaker, a long-nosed coupe with little trim and an air of motion about it, was acclaimed the top car of all time in a poll taken by the *Chicago Daily News* of stylists representing the Big Four automakers, General Motors, Ford, Chrysler, and American Motors." —*Automotive News*, May 1972.

First metal model of a new Studebaker design
intended for later mass production, 1962. Metal was
the next step after clay.

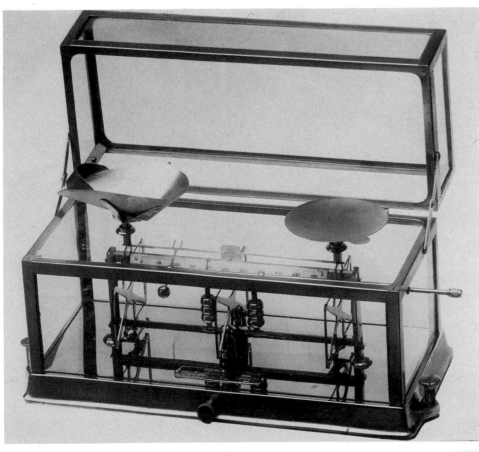

The logical design approach for the Torsion Balance was quickly recognized by the manufacturer and by users of such precision instruments. Its easy maintenance and protection from dust and dirt resulted in wide and sustained consumer acceptance.

All of our experimental designs were made of clay in 1/4-size models. The most promising ones were translated into full-size mock-ups, and, in order to save time and expense, we sculpted the two sides in two different versions. To experienced eyes, a visual dichotomy made it easy to appraise one side or the other as an entire car. Note: The reader can do the same thing by placing a sheet of paper over one side, thus hiding the other half.

Rosenthal china, 1940s.

Lounge of the *Loraymo* IV.

The *Brazil* and the *Argentina* leaving for South America, 1955.
We designed the interiors for both of these luxury liners.

Observation structure designed for the Moore McCormick Steamship Company, 1955.

The U.S.N. *Carronade,* designed in 1953 for the navy to improve the life of the crew in extreme conditions of warfare. Our goal was to limit noise and heat. We mostly worked on interiors, and some of our work is still considered "restricted."

Rotating light beams used as aircraft warning system atop the Empire State Building, 1955.

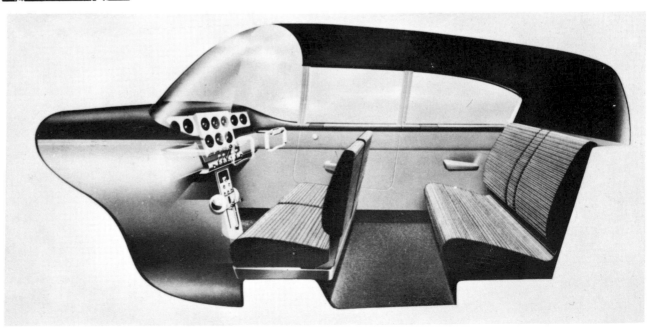

Interior of Fairchild Hillier helicopter, 1962.

Early and late rendering of Shell logo design, 1967. More on the Shell story appears on page 230.

was asked to improve the appearance of the Alouette, a French helicopter built by Sud Aviation.
ore starting our design research, I went to Le Bourget Airport with one of my designers for a test
ht with an ace pilot. My designer went first. The pilot gave him such a rough time with weird aerial
obatics that my colleague became ill. The pilot then asked me to get in and go up. Declining was
hinkable, so I went. I have never had such a flying experience. I have done all sorts of acrobatics
egular flights with air-force test pilots, but none of them compares with a wild helicopter ride.
eveloped a great respect—fast!—for the machine's versatility.

ouette after design of new fuselage, 1957.

About the same time as the Alouette, I tested a BMW and found it remarkably sprung and extremely powerful. I designed an entirely new body and had it built by my coach-builder friends, Pichon and Parat. The rear window provided excellent visibility and, like the Lancia Loraymo, the car had front-end protection. The two extended exhausts doubled as individual bumpers; made of thick steel plate, they could slide into their housing upon impact. Note the doors opening up into the roof, facilitating entrance and exit. The door openings also extended far forward, providing increased leg clearance. This car is now displayed in the Automotive Museum of the University of Southern California in Los Angeles. I first drove it from Turin to our farm La Cense, near Paris. Viola and I "baptized" it with champagne.

The BMW at La Cense.

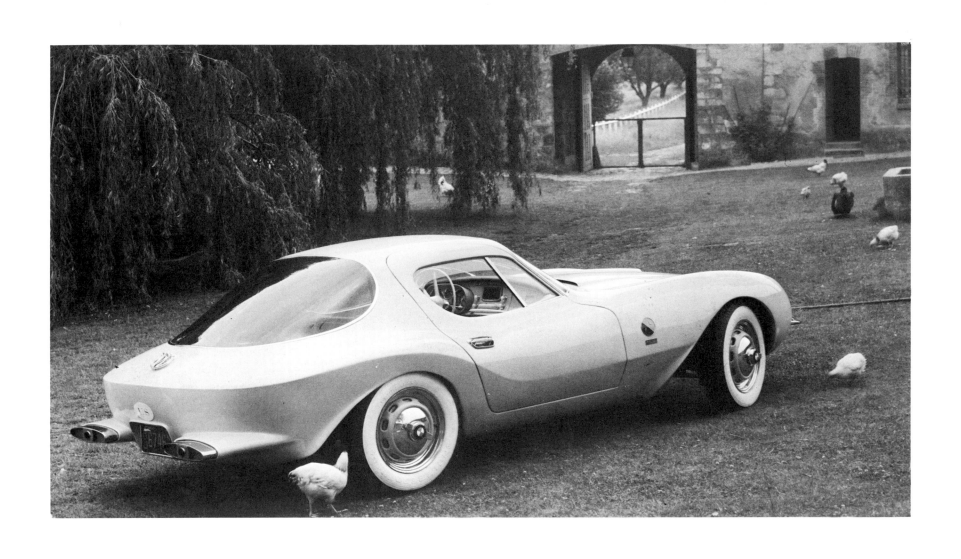

169

My next sports car was built by Boano in Turin on a Jaguar XK chassis. Its aerodynamics were excellent, as was its rear visibility. The wide rear-slanted pillar was noticed by automotive people. The press noticed, too, when a well-known German sports-car manufacturer adopted the same idea; the firm is still using the basic design idea. The XK is shown in front of the pigeon house at La Cense. The tail lights, turn-indicator lights, and brake lights were all incorporated in the rear bumper.

Three experimental designs; exercises in possibility. The cars were never built, nor were they intended to be.

Sliding doors are useful in countries where curbs are high, or where massive snowdrifts interfere with doors opening.

The car below would not be outmoded on a freeway today. Notice the extra-large speedometer on the dashboard. It had highly visible numerals and pointer, all well illuminated. Much present-day instrument-panel design is hardly readable. Stiff regulations as to the degree of visibility of all instruments is in order. Function and safety supercede all other considerations in styling a dashboard.

This design displays my early tendency for off-center hood-panel treatment, applied to Avanti in 1961.

The BP emblem before and after redesign, 1958, as applied to gas stations, tankers, delivery trucks, ships, containers, product packaging, road maps, aircraft, uniforms, tanks, refineries, letterheads, etc.

For more than fifteen years, C.E.I. has been retained by BP (British Petroleum) as design consultants for whatever carries its name. When the prototype station was tested and approved, we developed manuals that made building stations easy and precise, even in remote parts of the world. As a parallel system, we also developed a "Rehabilitation Manual" so that the features of the new prototype could be applied to most existing stations. This upgrading is a classic role for the industrial designer. Less than a week after our BP program was completed, C.E.I. signed an agreement with Shell International as design consultants. This long-term assignment covered an even larger scope of activities, including many for Shell Chemical.

My first impression of the existing BP logo was that it looked drab and that its colors were uninviting. As the logo obviously attracts the motorist's attention, its drabness has a negative cumulative effect on consumer awareness and, of course, on sales.

BP executives wished to keep the green and yellow colors in the form of a shield. I retained the services of Dartmouth University's color-research laboratory and asked them to determine the most visible green and yellow pigments under a wide variety of weather conditions. For color rendering of the BP logo, before and after, see page 218.

The U.S. Department of Transportation asked four automotive manufacturers to develop an experimental safety car. It would be tested for impact at fifty miles per hour in head-on collisions against concrete walls. Ford, General Motors, American Car and Foundry, and Chrysler were retained to build experimental vehicles. My involvement in 1972 stemmed from working with Fairchild on the body which was used on a Chrysler chassis. The films taken of the tests revealed that all the cars, except the Chrysler-Fairchild-Loewy unit, were smashed to bits. Our vehicle, shown here before and after the impact, suffered so little damage that the hood could be opened after collision and the rest of the body was practically untouched. Otherwise, the only damage was to our finances; due to a highly insufficient design fee, we lost fifty thousand dollars on the project, but we at least performed a service for safety.

One of the features of the well-known Lancia Loraymo was the airfoil angle, which could be adjusted to reduce the Kamm drag effect. The airfoil angle, in different forms and sizes, became standard in racing cars. When the Lancia Loraymo was displayed at the Paris Automobile Show, you almost couldn't see the car for the crowds. President de Gaulle was keenly interested in lowering gas consumption in France, and, when he came to see the Lancia Loraymo, we talked at length on the subject.

I designed an all-aluminum body and went to Turin, where a small coach builder built a prototype with excellent all-around visibility. The Lancia had good roadability and a low-hung chassis; it was quick and comfortable. An entirely new feature at the time was the elimination of the front bumper and its replacement with a heavy steel frame surrounding the grille. Fitting snugly inside the hood and mounted on coil springs, it could, in case of light collision, "give" slightly without damaging the body's aluminum sheathing. Special racing-car exhausts emitted a low-frequency rumbling sound, resembling that of a powerful sports boat. Mounted over the rear of the roof was an aerodynamic airfoil wing. I named the car Lancia Loraymo (Loraymo is my international cable address: *Lo-ewy Raymo*-nd).

For a trip to Europe with our young daughter, Laurence, we wanted a large, air-conditioned, chauffeur-driven car. I bought a Cadillac and shipped it to France. Its general appearance was acceptable, but for the front end and the extravagant rear fins. I redesigned the car and removed all the bulky chrome-plated extras, about a hundred and fifty pounds of junk. Nearly all the costs of the transformation were covered by selling the parts to a GM dealer. I retained the tail lights and incorporated them in the redesigned, simpler fenders, all aligned with a redesigned, simpler bumper. We lined a huge empty space in the front fenders ahead of the wheels with felt and so created two extra luggage compartments. The showy wheels were simplified with Avanti hub caps. In the cowl, immediately ahead of the windshield, we provided large circular openings for better ventilation. The color, like most of my cars, was a light metallic beige. For those who did not mind driving a chrome-plated barge in traffic, the standard Cadillac of the period was a comfortable, excellent machine.

I often speak to young, prospective automotive designers. In an attempt to wean them from ostentatious display often prevalent in Detroit, and back to clean design, I show students these slide projections. Over the years I took hundreds of snapshots of parked cars with monstrous fins and grilles. The unfortunate poodle serves as a yardstick, and the point is effectively made—the audience always laughs. The French say, "ridicule kills," but Detroit nevertheless survived the large-fin era.

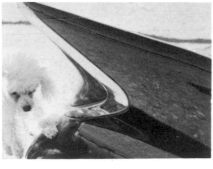

Avanti

The birth of Avanti is a short, happy story. Sherwood Egbert, president of Studebaker, phoned me in Palm Springs early in March 1961 to ask if I would design a sports car. I said yes and flew to South Bend. On March 6, with Gene Hardig present, I got the assignment: The car had to be built on an existing chassis, and Gene gave me a full-size blueprint to take back with me to California. Although Egbert promoted the sports-car idea for a specific market, he never actually had a specific concept in mind, and he never showed any conceptual sketches to Gene or me. What he was was a masterful manager with an imaginative mind. On March 9, and for the next few days, I began work in my studio on diagrammatic sketches for the car I envisioned. I provided side elevations, front

and rear views, as well as a horizontal projection, and I mounted the group on four 36 x 18–inch cardboard panels. Then I rented a rather run-down, two-room building in the desert where the three designers I had selected as project colleagues could work and sleep. Rented drafting tables were quickly installed and the wood and clay purchased. When the team assembled on March 19, the conceptual panels were taped to the walls, and I explained the concept in terms of Studebaker's mandate and my design notion. We began work early the next morning.

In order to save time, I placed Johnny Ebstein, an excellent designer and organizer, in charge. His task was to keep his two associates on the basic outline as a targeted design

I had just as difficult a time convincing the photographer who took this picture to leave ut the background as I had when the Hupmobile was shot, thirty years earlier. It wasn't asy to finally achieve the simple effect of highlights and shadows, which expressed the vanti style.

al. Seven days later, a ¼-size clay model and perspective renderings were ready, and flew to South Bend to show them to Egbert. He and Gene Hardig gave their approval, nd I flew back to Palm Springs to complete work on the detailing. Piloting his own small rcraft on April 2, Egbert arrived in Palm Springs, liked what he saw, stayed only an our, and flew back to South Bend. Two days later I flew to South Bend, where work began n a full-size clay mock-up; this was fifteen days from the project's inception. I supervised development closely, modeling some areas myself; subtle accents could not be explained ny other way. I often appraised the "flow" of Avanti by closing my eyes and running my ands over its body. On April 27, we presented it to the Studebaker board, Egbert and

Hardig present, of course. It was immediately accepted, encouraging my notions of the aesthetics and courage of at least some financial types.

The Avanti story has become well-known in the automotive industry, not only because the car soon achieved classic status, but because it was developed so efficiently; a lightninglike shoe-string operation compared to the hundreds of thousands of man hours and millions of dollars Detroit expends on an average new body design to achieve a "committee" look. With Avanti a great part of the expenses were for airfares between California and Indiana, plus a few cases of champagne at appropriate moments.

Things have changed, and now Detroit production is often well designed, if not entirely individual; but it still costs millions to develop a new body design. Ten years ago, Daniel Jedlicka wrote an article in *Esquire* called "Instant Classics." He included ten cars, two of them Studebakers. "$2,500 will get you a 1963 Studebaker Avanti," he wrote, *Raymond Loewy styled it and liked it even better than his slick '53 Studebaker coupe. Ian Fleming bought one. The roof was trimmed with a steel boxlike frame attached to a hefty roll bar and windshield support. There were air-craft type rocker switches mounted in the roof, a Paxton supercharged V-8 and a fiberglass body with tremendous impact resistance. It hit 170 m.p.h. at Bonneville. By 1975 it should be worth $7,000.*

Today, it is worth far more, and hard to find at any price. Jedlicka's comments about the Studebaker Hawk were in the same vein. The Avanti was introduced at the New York Automobile Show in 1962, and it caused great comment. *Product Engineering*, in its June 1962 issue, noted:

Introduced in New York during April of this year, the Avanti was test-driven five days later on the Studebaker testing grounds in South Bend by Product Engineering staff member, J. J. Kelleher. He drove the car, which was equipped with a four-speed manual shift, at all speeds (up to 130 m.p.h.) and in all the driving conditions the test track afforded. This test included emergency stops from 100 m.p.h., with two wheels on gravel, two on regular paving. The car stopped within 450 feet without swerving, brake fade, or excessive nose dive. Kelleher, who managed to equal acceleration times set up by Loewy and Egbert as a performance goal, reports that the car's chassis is the most rigid and stable he has ever driven.

An officially recorded test performance on Salt Lake City flats was certified by the U.S. Auto Club, which noted: "A supercharged Avanti posted 32 new records—for a tremendous total of 337 new records. To top it all off, an experimental Avanti made a one-way run certified at 196.62 m.p.h."

I still keep two beige Avantis, one in Paris, the other in Palm Springs.

Inside our desert workshop and visible behind me, taped on the rear wall, are some of the diagrammatic charts prepared before the design team arrived from the East. Time was short, and I emphasized the main requirements: minimize chrome; avoid decorative moldings; accent the wedge-shaped silhouette; stress long, down-slanted hood; abbreviate the rear and tuck it under; place instrument panel overhead, above windshield as in aircraft; install aircraft-type knobs and levers on the console; pinch the waistline, as Le Mans-type racing cars; design hoods with an off-center panel; accent spacecraft "reentry curve" wheel openings; simple disc wheels; above all, think aerodynamics.

Thirty rough sketches concept; indicated the design target for the team members. This system kept us on a true course, avoiding repetition and detours.

There was no time for several scale clay models; we built only one. As soon as a shape was completed, we took snapshots for further evaluation and quickly tried another version on for size. All these are the same clay model—sometimes modified two or three times a day. To make the clay easier to use, we kept it warm outside in the sun—every minute counted.

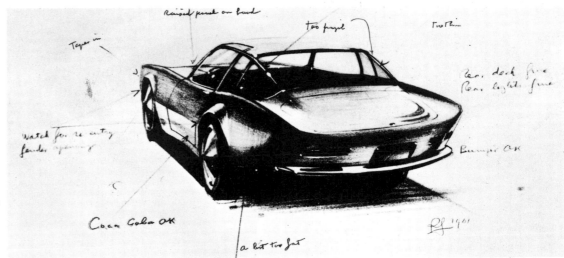

An idea for a two-seater roadster that was never built. The pinched waistline of Avanti, which I nicknamed the "Coca-Cola countour." Most racing cars used it at Le Mans—for instance, Jaguar, Ferrari, Aston Martin, Mercedes, and Lotus. The inspiration started there.

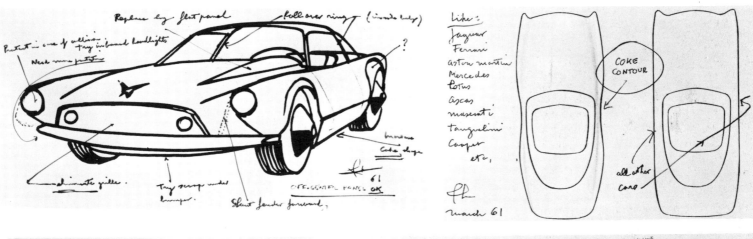

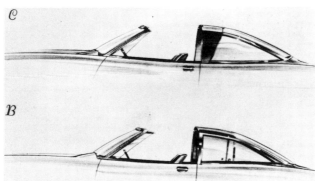

Some ideas for incorporating an easily removable panel over the front seats. Storage space was provided in the trunk.

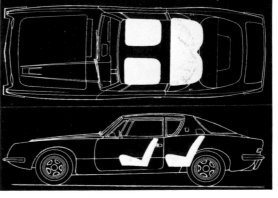

Avanti's interior layout. Notice behind the back seats a removable panel providing access to the trunk. The plan view illustrates the hood's off-center raised panel parallel to the chassis center line. In line with the steering wheel, the effect during long drives is orderly and relaxing.

The model team working on the full-size mock-up from the clay model (shown by arrow) we flew in from California.

Gene Hardig and Raymond Loewy confer with the model team.

The full-size model of the passenger steering and control layout.

This snapshot was taken after the owner of an Avanti, driving at high speed on a California freeway, died of a heart attack at the wheel. The car turned over several times, but the built-in roll bar remained intact. The driver had no passengers, but they might well have survived wearing seat belts.

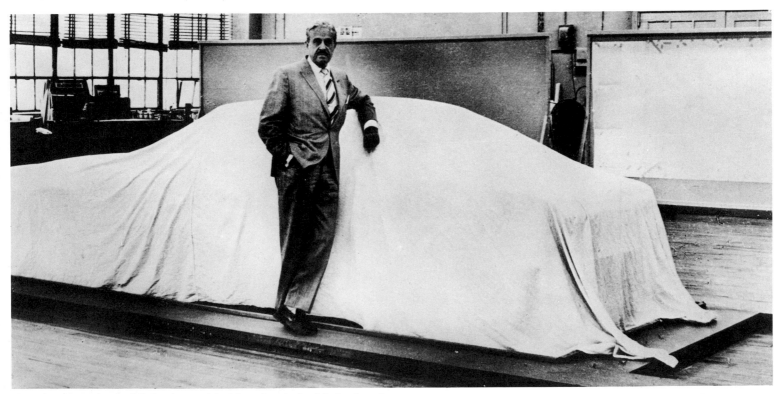

I was ready to display the full-size clay model of Avanti to the Studebaker board.

The completed full-size clay mock-up.

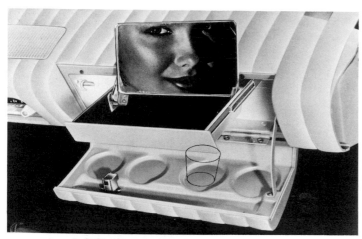

I designed a compact vanity for the 1951 Studebaker, sliding out of the dashboard on the passenger side. Detroit adopted this some twenty years later for its luxury models.

The full-size clay mock-up completed. Notice the two wide exhaust outlets suggested but not adopted.

These cars convey the impression of speed, even when at rest.

Avanti 1961

Raymond Loewy

When the Avanti was shown at the New York Automobile Show, it attracted attention and a slew of orders. South Bend could not supp[ly] car fast enough; the fiberglass supplier had never fabricated fiberglass bodies in large quantities. Buyers had to wait so long that they began to cancel their orders. This, at least, is the story we were told by the Studeb[aker] people. Avanti was soon in trouble, and even later sales were badly affected. After Studebaker went out of the car business, a corporation w[as] formed just to make Avantis. The cars are assembled by hand, each auto-mobile virtually a custom job. Gene Hardig is still on the scene, and Ava[nti] sales continue to increase. New manufacturing facilities are now being b[uilt]. The engines today are of GM manufacture, installed by Avanti's dedicate[d] craftsmen.

This snapshot, taken in Moscow in 1974, shows the handwork of a young Russian automotive fan—very likely inspired by the Avanti. He built the vehicle by placing two motorcycles in tandem, and the raised panel on top of the hood accommodates the handlebar. The driver rides on the bike's saddle. The body color even closely approximates the soft red of Avanti's first production models.

In front of the industrial design exposition at the Louvre.

Dr. Djerman Gvishiani and I in front of the Plaza-Athenée Hotel in Paris, following the signature of a protocol of cooperation. Dr. Gvishiani, brother-in-law of Alexei Kosygin, is chairman of the U.S.S.R. Council of Ministers' Committee for Science and Technology. According to the protocol, I was to become the industrial design consultant to the Soviet Union, advising on all mass-manufactured products.

The Avanti being carried inside the Smithsonian Institution across from the White House in Washington. It remained on display from August to November of 1975. Also displayed were the original Ford Model T and a few other historical automobiles of the 1900s: Avanti was the only contemporary automobile shown.

If I were to redesign Avanti today, I would keep it much the same. I might, however, emphasize the off-center treatment and place two fog lights into the headlight cluster. The body could be a bit wider and lower, designed to accept wider tires. The face of the instrument panel would be white and the pointers black for better readability. The exhaust system needs improvement. The windshield wipers could disappear under the cowl. Better sun visors are essential. The air-cooling system needs study. The engine compartment air-intake remains where it is but could be redesigned for better efficiency. A spoiler might be desirable. I would reduce the width of the A pillar (front windshield pillar) to a minimum to reduce blind area. The French Citroen A pillar is sturdy but far thinner, a good example of safe design.

Study for post-Avanti sports car, with airfoil deflector for fuel economy.

Studebaker that was never seen, 1962.

I picked up this student's sketch in 1960 while visiting an industrial ▶ design school where students who wish to become automobile-body designers were being trained. Unfortunately, unfettered expression without a technical base is of little help to student or society. After graduation and on a first job for an automobile manufacturer, the former student is suddenly confronted with the tough realities of designing four-door family sedans at rock-bottom cost. Understandably, he feels helpless, unprepared, and frustrated.

In an effort to guide students into usable design directions, I pointed out the following problems in the sketch.
1. The door is so enormous and heavy that it cannot be hinged, nor can the window be opened or closed. In town the door would hit the curb.
2. Due to the position of the seats, the driver cannot see the road ahead.
3. The sharp angle of the windshield is impractical: due to the large glass area, people in the front seats driving in hot weather or stalled in summer traffic would probably be broiled.
4. There is no way to change the rear tires.
5. There is no bumper protection: a rear-end traffic collision would smash the body and probably the passengers.

Teachers should pass on a sense of technical responsibility; freedom of expression should be oriented toward utility.

These three basic styling concepts underpin most body diagrams developed in automotive-design studios. The diagrammatic drawings were made to illustrate a point.

In the top case, there seems to be conflict between forward and rear motion.

In the center approach, favored by this designer, it might be considered that, in a getaway, gravel would be projected at an angle and in a direction consistent with the general silhouette.

In the bottom sketch, the effect is immobility. This is the approach used for the highly popular Mercedes Benz, technically one of the best cars made. Its highly conservative look, almost an anachronism, apparently appeals to many; Detroit has brazenly copied this shoebox approach.

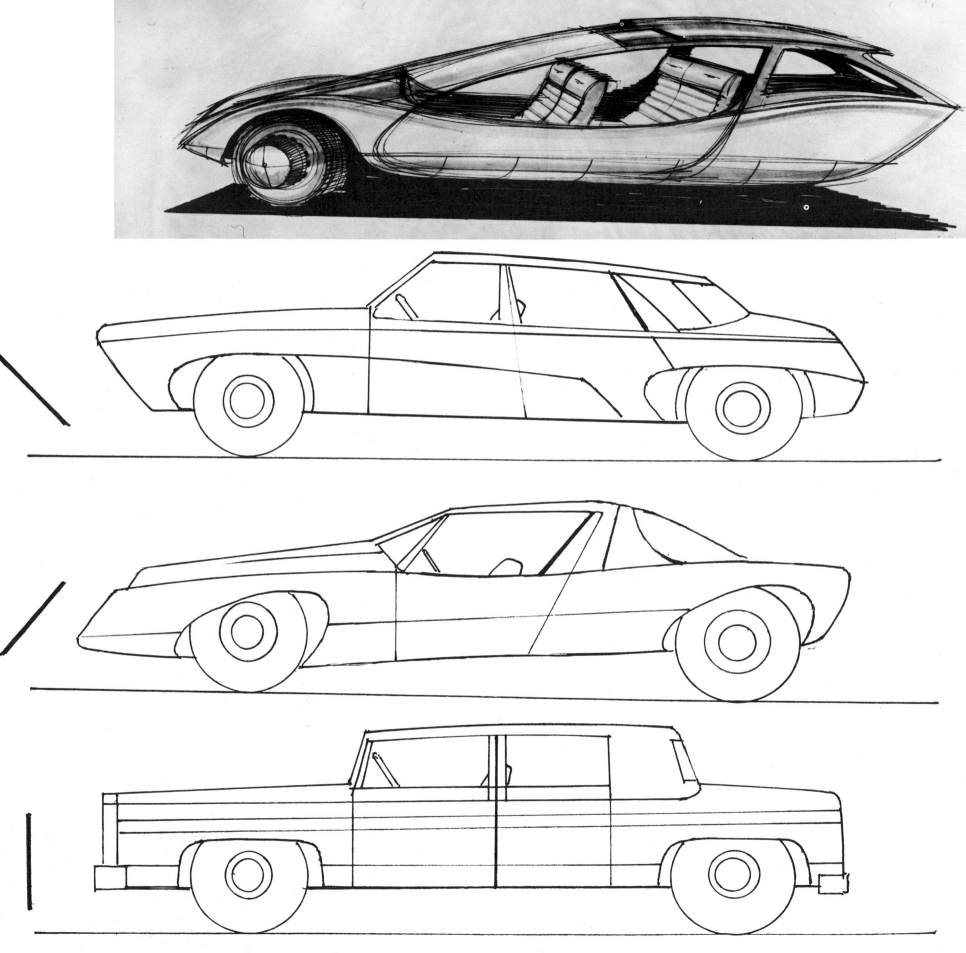

I designed this decanter for Brown Forman in 1958, as a gift package. It has been widely copied ever since by manufacturers of cosmetics, medical compounds, furniture polish, rat poison, and sugar substitutes.

The copy was written by Brown Forman's public-relations department. What is interesting here is the way a product was sold in the 1950s, using the designer's name.

$\mathcal{S}ee$ the Decanter Sensation of the Year— Designed by World-Famous

Raymond Loewy

Old Forester

I was in my New York office in March 1966 when I received a visit from Jersey Standard's counsel; he had come, he said, at the suggestion of the board of directors. The company had decided to change the name Esso into another one, and, for reasons he chose not to explain, the fact was to be kept in the most severe conditions of secrecy. the project was to be known as "Nugget."

We signed a contract and I left immediately for my winter residence in Palm Springs, California.

In less than a week I found the name I wanted: Exxon, and I made seventy-six rough pencil sketches based on the word, placing the visual emphasis on the double x's.

After making photocopies for my files, I air-mailed the originals to Joseph Lovelace, then vice-president of my company in New York, whom I had selected as Nugget's project manager.

I indicated the version of Exxon which I preferred, and it was eventually selected, adopted, and used by Jersey Standard. The sketch shown here is the original one I made in Palm Springs in March 1966.

Then a large team of Jersey Standard's executives started—in close collaboration with my team—the long research, explorations, and so on, essential to such a major transition as a change of name and logotype for one of the world's largest corporations. The end result of such intimate cooperation, carried out in top-secret condition, was the word *Exxon*.

I valued the double x for its neo-subliminal memory-retention value and also for a certain similarity to the two s's in Esso.

march 1966
Palm Springs - Cal

In our New York office we maintained a fully equipped shop in which full-size models could be built, such as this Fairchild Hillier helicopter. At right, Fred Toerge, a talented, imaginative designer who became my assistant in charge of the *Saturn-Apollo* application program and the *Skylab* projects under my direction.

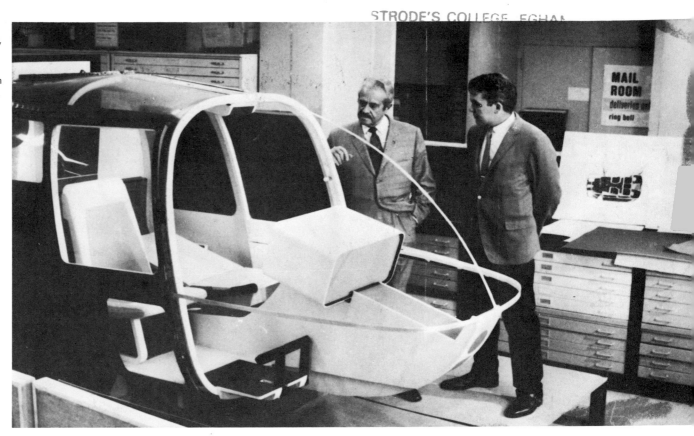

The Shell emblem is called a pecten, the name for this sort of shell. The original sign, when viewed from a distance or in unfavorable light, was somewhat obscured by the firm's name. When we redesigned the pecten, our main concern was to augment its visibility by a) removing the firm name and placing it adjacent to and under the pecten, on a rectangular white panel, b) selecting a more brilliant yellow and red, and c) placing the pecten, redesigned with thinner ribs, against a bright red square, or outlined by a wide red band.

The new logo was tested in various parts of the world and found effective. It is now used internationally in 130,000 stations on all Shell items. After our work for Shell was completed, New Jersey Standard asked us to work on the new name and logo for Esso.

The uniforms C.E.I. developed for Shell were carefully designed from both male and female viewpoints, as well as for all weather conditions. Comfortable and fireproof, the uniforms are unaffected by fuels and are easy to keep clean. We discovered that many employees were quite style conscious. The turned-up cap visor proved a popular feature. Shell liked our prototypes and they were first adopted in Europe. The male model happens to be the managing director of C.E.I.; the young lady is one of the secretaries.

The interior of the famous Dassault Mystere executive jet. We designed an especially luxurious one for the Aga Khan.

NAVY SHIP FURNITURE

EXISTING   **PROPOSED**

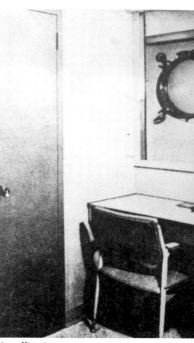

An officer's stateroom.

In 1962 I made a proposal to Cyrus Vance when he was secretary of the U.S. Navy, showing ways to improve crew comfort and save money at the same time. Previously we had established new standards of comfort and work efficiency for U.S. Navy warships and also for the Merchant Marine Commission. We also designed the interior appointments of the experimental Navy submarine *Tektite*, as well as the first nuclear destroyer. Upon surfacing and debriefing, the captain attributed much of the success of the operation to the efficient ambiance.

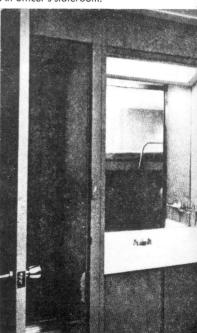

This Dictaphone won first prize in an industrial design competition in 1965 for mass-manufactured products.

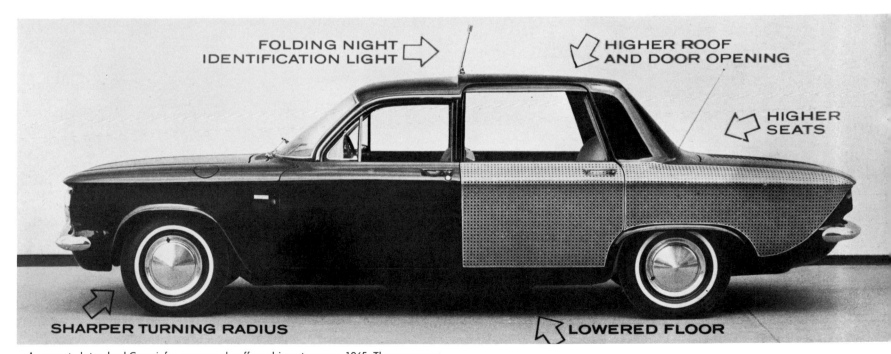

FOLDING NIGHT IDENTIFICATION LIGHT ➡

⬇ HIGHER ROOF AND DOOR OPENING

◀ HIGHER SEATS

◣ SHARPER TURNING RADIUS

◤ LOWERED FLOOR

A converted standard Corvair for use as a chauffeur-driven towncar, 1965. The passenger compartment was made more roomy by lowering the floor, and the steering engineering was modifed for a smaller turning radius.

Frank Lloyd Wright and I shared a hatred for telephone poles and power lines, an aesthetic visual rape inflicted upon nature. Economically acceptable in some areas, they are inexcusable in others, and I hope that urban planners in the future will make them an issue of high priority. Having found this depressing painting in Paris, I sent a color photo of it to Wright, who invited me to come and spend a few days in Taliesin West. We consoled each other on this and many other societal issues which we felt powerless to change.

NASA

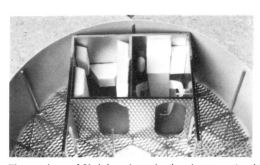

The mock-up of *Skylab* as I saw it when I was retained as habitability consultant.

Dr. George Mueller and Raymond Loewy at the decisive 1968 NASA conference in Washington, D.C.

Eight habitability studies comprising over eight hundred pages and over a thousand illustrations, statistical analyses, diagrams, and scale models were prepared during a six-year period by a team under my direction. Headquarters were in New York City, but work went on in Texas, Alabama, California, and other states. The research and development related to both *Skylab* and the earth-orbiter shuttle's first conceptual studies.

In order to appraise the value of the efforts, I quote a passage from a letter by Dr. George Mueller, NASA's deputy administrator for manned space flight:

Dear Raymond:

Two significant events in the last decade will, I believe, shape the future course of human history. The first was landing on the moon with its demonstration that humanity was no longer bound to the earth. The second was the manned orbiting space station with its demonstration that man could live for indefinite periods of time in a weightless environment and that he could perform useful, yes, unique work in that environment.

Raymond, in my opinion, you and your organization played a crucial role in the latest of these momentous steps that man is taking to the stars. I do not believe that it would have been possible for the Skylab crews to have lived in relative comfort, excellent spirits, and outstanding efficiency had it not been for your creative design, based on a deep understanding of human needs, of the interior environment of Skylab and the human engineering of the equipment and furnishings which the astronauts used. That design and engineering applied, in turn, to our follow-on space stations has provided the foundation for man's next great step—an expedition to the planets.

You should be proud, as all of us who know of your contribution are proud, of the key role you have played in laying the foundation for man to live in space.

My most sincere congratulations for your work on the Skylab program and my best wishes for your continued contributions to man's role in space.

I received this letter on July 31, 1974. Earlier I mentioned my introduction to NASA, courtesy of John F. Kennedy in 1962. I established a relationship with NASA administrator Jim Webb and his assistants at that time; but it was not until 1967 that I received a call from Denver: Would I be interested in flying there to discuss the possible employment of the Loewy organization as NASA's habitability consultants, "to help insure the psycho-physiological safety and comfort of the astronauts." From our own offices in New York we would work with the engineers, scientists, and space-medicine specialists in charge of the project. The assignment was called then the Earth Orbital Workshop (EOW). Twelve years ago, remember, little was known about the possibilities of human survival in exotic conditions of zero gravity, where man might be subjected to strange forms of space sickness, showers of asteroids, etc. I was a bit staggered when told that I was to help create a space vehicle in which two or more men could go through lift-off, perform up to three months in space, survive, and reenter earth's atmosphere safely. I accepted instantly.

After an extensive indoctrination in space matters, learning fascinating space jargon (which I called "orbit-talk"), and whirlwind visits to Houston, Texas; Huntsville, Alabama; Huntington Beach, California; and St. Louis, Missouri, I let myself be "brainwashed" into the zero-gravity feel. My assistant was my friend, Fred Toerge, and, at the suggestion of two astronauts, we put on spacesuits and went through the routine astronaut tests. A space-medicine medical officer at Wernher von Braun's installation in Huntsville, Alabama, supervised us, and we quickly learned by experience that any kind of motion while wearing a pressurized suit is exceedingly strenuous, a valuable lesson for our later conceptual efforts.

I formed a team of about a dozen bright young industrial designers and we immediately set about the task. Due to the understandable lack of data regarding the problematical chances of prolonged survival in deep space, we frequently relied upon logic and educated intuition. The project became known as Skylab, and I spent many sleepless nights on it during the next few years. My personal responsibility oppressed me and I prayed that our basic concepts were right. I was blessed with the support of an outstanding team.

The pressure was partly relieved when *Skylab One* completed a twenty-eight day mission. Astronaut Paul Weitz was aboard and he, like Jack Lousma, had helped me convince top NASA management and other astronauts that the inclusion of a porthole was essential. I stated that I could not fully endorse a capsule in which there would be no possibility for months at a time to look out and see our earth. All the *Skylab* crews during debriefing stated that without the porthole the mission might have been aborted. They recommended that more and larger portholes be included in all future spacecraft. Our team made three other psychological recommendations; adopted, they were critical to Skylab's success: First, we recommended that each astronaut be allowed eight hours of solitude daily. Second, astronauts were secured for meals facing each other, as they would be on earth. Lastly, all partitions were to be smooth and flush to make it easy to keep them clean in the event of uncontrollable space sickness. At the time, NASA was divided on a number of philosophical points, and it was Dr. George Mueller who backed me up. It was he who organized the special session in 1968 at NASA's Washington headquarters during which I could explain my position and answer questions. Finally, those who were in favor of my theories prevailed; without Dr. Mueller's vision and common sense at a critical moment, they and I would have lost on significant issues.

On July 2, 1970, Caldwell Johnson, chief of the Spacecraft Design Office in Houston, sent a report about our performance to NASA headquarters. I quote some excerpts:

The Loewy Company has been providing support to the Habitability Section of the Spacecraft Design Office since September 1, 1969. Their prime contribution was to provide advice and counsel in interior designs, drawings and renderings of concepts and models and mock-ups. Also, their support was to include the Skylab, Shuttle, Space Station and Base projects. In performance of their task, I have listed some of their accomplishments:

A detailed list of eight major projects followed, as did this conclusion:

Loewy personnel constructed mock-ups of the Skylab hygienic facility and sleep station at Manned Space Craft as a part of a presentation for top management review. The mock-ups were exceedingly well done and in a very short time period. The mock-ups were well received and served to highlight specific aspects of habitability which are best understood in three dimensions.

The above tasks are essentially in chronological order and as indicated cover a wide range of design areas and projects. Not included in the above comments are the numerous models which were submitted prior to some of the full scale mock-up activities.

The support given to the Spacecraft Design Office has been enthusiastic, competent and cooperative. Perhaps the single most outstanding feature of the Loewy Company is their ability to accomplish these tasks in surprisingly short periods of time. It is a credit to their ingenuity and management. In conclusion, it is fair to say that we have received sound, professional support from the Loewy Company

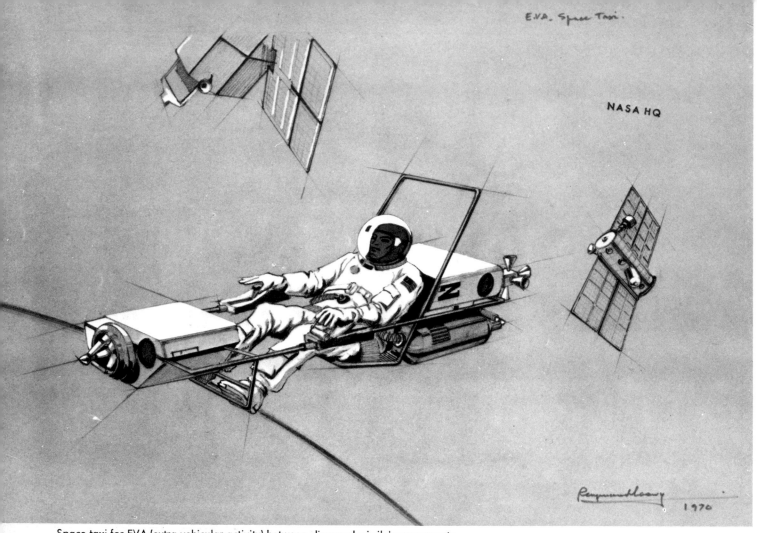

E.V.A. Space Taxi.

NASA HQ

Space taxi for EVA (extra-vehicular activity) between dispersed missile's components.

Volumetric three-dimensional studies for an artificial gravity, nuclear-powered space station.

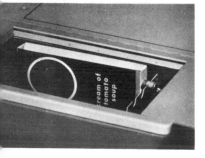

Studies made in our New York office of experimental dehydrated foods, in this case, tomato soup supplied to us by the research-and-development division of NASA. Illustration shows the soup container in its receptacle, with hot water injected into tomato-soup mixture, and later use.

Study of device facilitating the transfer from working overalls into EVA spacesuit before leaving *Skylab* through hatch.

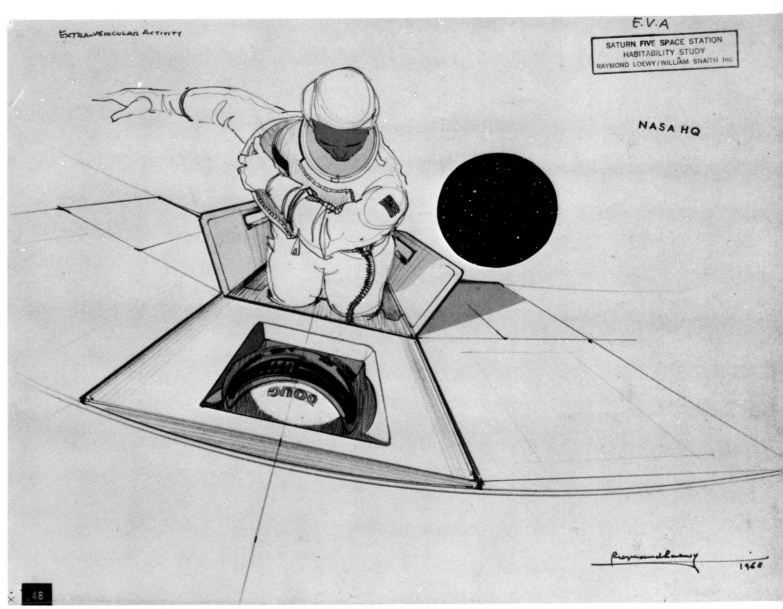

EXTRA-VEHICULAR ACTIVITY

E.V.A

SATURN FIVE SPACE STATION
HABITABILITY STUDY
RAYMOND LOEWY / WILLIAM SNAITH Inc.

NASA HQ

Raymond Loewy 1968

Inside the *Skylab* mess in front of the food unit at left, the porthole that played such a large part in the mission.

Notice various storage units in background, built absolutely flush with partition and with high-polish, easy-to-clean surface. This proved highly efficient when space sickness caused vomiting, making it easier to remove floating food fragments in zero gravity. The food cluster is set up in a triangular layout; I was opposed to preferential hierarchical positions for crew members during long missions.

The articulated plywood dummy representing a cross section of the astronaut's physical characteristics, often used for quick reference.

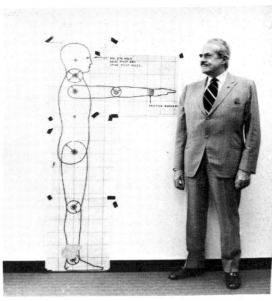

Then a further report dated August 30, 1971, to NASA's executives in Washington:
In recent discussion with Spacecraft Design Division people, you requested our appraisal of the support effort provided us by Loewy, Inc., under subject contract.

In the past, we have commented on some specific contributions from Loewy, Inc., which have improved the quality of our designs and concepts for Skylab and for the Shuttle-launched Space Station. They have continued to perform well in that capacity, and are now helping us with the Shuttle.

Loewy, Inc., has been especially useful by helping us to reduce the harsh appearance of some of our design concepts while maintaining their functionality. In some instances, Loewy, Inc., has significantly improved the concept from an engineering standpoint. Their treatment of design problems is very fundamental; and, if the requirements are properly stated, their designs are usually functional and simple.

The company has developed a unique capability to translate concept sketches to 3-dimensional mock-ups in extremely short times, and at a lower cost than we can obtain from other sources. The quality of their products is also excellent.

Both the concept design and fabrication of 3-dimensional engineering tools is done with a minimum of direction and the cooperation of the company is excellent. Our close working relationship and mutual understanding of space related habitability problems has eliminated the requirement for time-consuming paperwork. This in itself allows our engineers more time for technical problems.

In summary, we feel that Loewy, Inc., has performed in a superior manner, and we look forward to their continued support.

Naturally, all at Loewy-Snaith were pleased with the kudos. But my personal pleasure related to the profession of industrial design. It was a long way from the Gestetner copying machine to NASA and Skylab. And there is an even greater future for the industrial designers of tomorrow.

On July 2, 1970, Caldwell Johnson, chief of the Spacecraft Design Office in Houston, sent a report about our performance to NASA headquarters. I quote some excerpts:
The Loewy Company has been providing support to the Habitability Section of the Spacecraft Design Office since September 1, 1969. Their prime contribution was to provide advice and counsel in interior designs, drawings and renderings of concepts and models and mock-ups. Also, their support was to include the Skylab, Shuttle, Space Station and Base projects. In performance of their task, I have listed some of their accomplishments:
A detailed list of eight major projects followed, as did this conclusion:
Loewy personnel constructed mock-ups of the Skylab hygienic facility and sleep station at Manned Space Craft as a part of a presentation for top management review. The mock-ups were exceedingly well done and in a very short time period. The mock-ups were well received and served to highlight specific aspects of habitability which are best understood in three dimensions.

The above tasks are essentially in chronological order and as indicated cover a wide range of design areas and projects. Not included in the above comments are the numerous models which were submitted prior to some of the full scale mock-up activities.

The support given to the Spacecraft Design Office has been enthusiastic, competent and cooperative. Perhaps the single most outstanding feature of the Loewy Company is their ability to accomplish these tasks in surprisingly short periods of time. It is a credit to their ingenuity and management. In conclusion, it is fair to say that we have received sound, professional support from the Loewy Company

A few of the reports supplied by our team to NASA. The information was precise and the reports were abundantly illustrated with diagrams, charts, mechanical drawings, and perspective sketches.

Orbiter Crew Compartment/X-Axis Docking

Allen J. Louviere

ry 1972

er operation in 3 modes of orientation - launch, Zero-G, re-entry.

crew -maximum of 4 crewmen occupy crew compartment at launch.

ompartment consisting of a 94"H X 108"W X 196"L module.

ate access necessary to emergency escape hatch and head during launch.

assage through the crew compartment extending from the airlock to the rear
d.

elop a working crew compartment configuration which satisfies the operational
ments for launch, Zero-G and re-entry modes.

ng the task assignment, a meeting was scheduled for January 12, 1972.

lly, Raymond Loewy/William Snaith, Inc. felt that the Shuttle Orbiter 040A Crew
tment/X-Axis docking failed to coordinate the launch, Zero-G, re-entry orientation
into well defined activity areas. Specific criticism included:

entrance to the Hygiene Compartment is difficult. Extreme body maneuvering
ired to gain entrance.

stems engineer has no visual contact with the pilots.

to the emergency escape hatch is awkward. Unnecessary maneuvering is required
h exit platform.

5 and 6 store for launch which require undue preflight and flight adjustments.

y through the craft is very limited in all flight modes. A specific passageway
be established to organize flow through the vehicle.

s is dispersed throughout the vehicle. A more integrated plan is needed.

Compartment Description

sis was placed on the accessibility of the emergency escape routes during launch
ilability of the Hygiene Compartment to individual crew couches. An aisle was
shed to define routes of travel and outline activity areas. In the Crew Compart-
Figure A4), the head and galley are along a common wall with a 20" passageway
n the wall position and the structural supports of the space couch.

vantages

The crew compartment has been reorganized to socially orient the space couches
to allow the entire crew to have visual contact with each other. The couches
can be tilted to suit individual preferences.

PREPARED FOR **NASA** BY
RAYMOND LOEWY/WILLIAM SNAITH, INC.
110 EAST 59 STREET, NEW YORK, N.Y. 10022

FOREWORD This report covers work accomplished by Raymond Loewy/William Snaith, Inc.
for the Manned Spacecraft Center, Houston, Texas, from January 24, 1972 to
January 27, 1973.

As a consultant design office, we have provided habitability design services
for the Shuttle Orbiter Program as requested by MSC. It was our goal to
quickly provide NASA with a variety of creative solutions for the stated
tasks. Sketches, mock-ups, mechanicals and models were the mediums
used to communicate our ideas and to establish a foundation for future
development.

Raymond Loewy
Project Director.

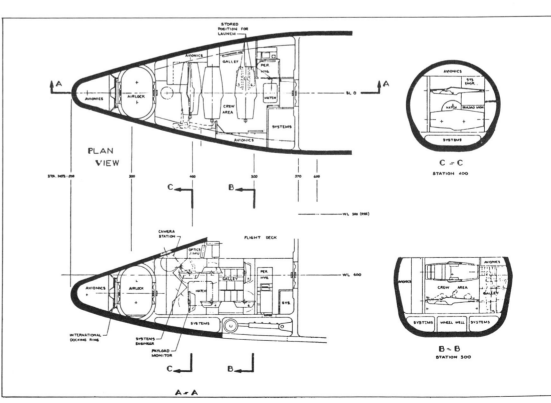

A1 - MSC Shuttle Orbiter Crew Compartment/X-Axis Docking

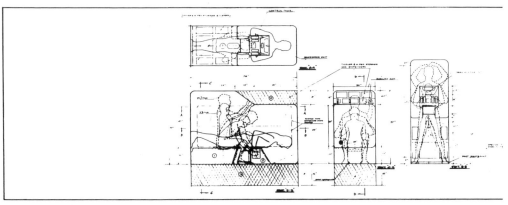

C12 Loewy/Snaith Approach A --- Mechanical

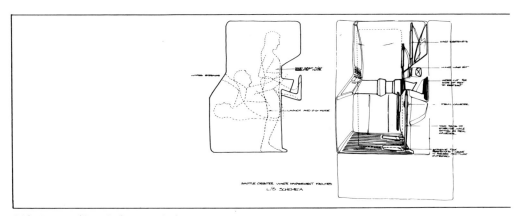

C13 Loewy/Snaith Approach A ---Perspective

All mobility aids and restraints are built in or flush with the adjacent surfaces in the area in which they are located. This approach required the minimum total volume of all concepts studied with no reduction in maneuvering ease.

Figure C14 illustrates Approach A integrated into an MSC, Houston scheme, allowing the vehicle wall curvature to intrude into the compartment without creating a headroom clearance problem previously experienced. Oriented in the shuttle thusly, the handwash unit is now accessible in all orientations.

APPROACH B
FIGURES C15-C16

This arrangement is somewhat similar to the NASA arrangement; however, by turning the fecal/urinal collector 180°, as shown in Figures C15 and C16, the headroom clearance problem for a person using the urinal/fecal collector is eliminated.

A second urinal opening, leading to the same collector tank, was placed below the fecal opening for standing urination use only.

Mobility aids and restraint provisions are built in or flush with surrounding to the maximum extent possible.

APPROACH C
FIGURES C17-C18

This scheme positions the personal hygiene unit and fecal/urinal collector perpendicular to axis of shuttle orbiter. The use of waste management units in this orientation varies according to the attitude of orbiter.

1   Launch - Fecal/urinal collector used in seated position. Handwash could be used by side access. Ingress would be accomplished by stepping down through the door into the foot restraint of the fecal/urinal collector.

2   Zero-G-All systems in use.

3   1-G Flights - Handwash and urine collector are usable. Ingress is accomplished by walking vertically through door.

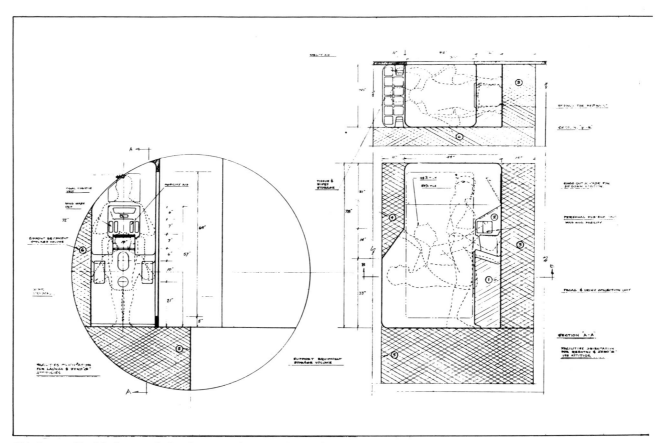

C14 Loewy/Snaith Approach A - Integrated into a MSC Scheme

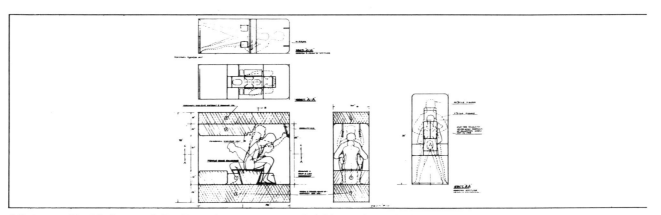

C15 Loewy/Snaith Approach B - Nasa Arrangement with Additional Head Clearance

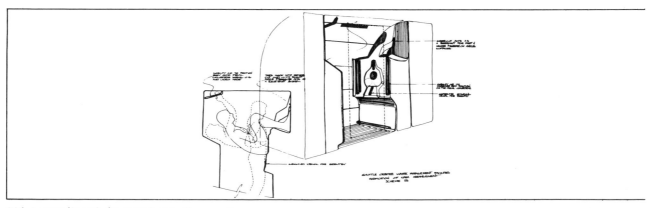

C16 Loewy/Snaith Approach B in Perspective

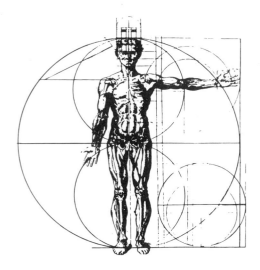

FINAL REPORT
VOLUME II OF TWO VOLUMES

HABITABILITY STUDY
EARTH ORBITAL SPACE STATIONS

PREPARED FOR NASA

BY

RAYMOND LOEWY / WILLIAM SNAITH, INC.

425 PARK AVENUE · NEW YORK · N.Y. 10022

Figure 55 is a photograph of a rough cardboard full scale mock-up we constructed of the approved WM concept. The fecal collector is configured to accommodate the body in a "Lean On" posture. Directly in front of the collector are storage compartments and the processor unit, readily accessible to the occupant tethered on the fecal collector. The defined quarter circle area in the front of the compartment represents space devoted to a body wash or shower stall.

As mentioned earlier, it is our recommendation that all equipment and stores necessary for fecal collection and processing be accessible and operable from a tethered position on the fecal collector. In Figure 61 we schematically illustrate how the bag described in Figure 60 could be installed and removed from the fecal collector while the occupant remains tethered. We suggest that the receptacle for the bag be attached to the collector with parallel double hinged arms allowing the receptacle to be swung down from the buttocks and following an arc outward to a loading position between the occupant's knees.

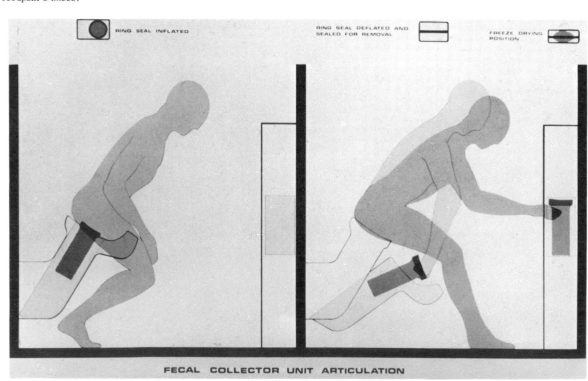

FECAL COLLECTOR UNIT ARTICULATION

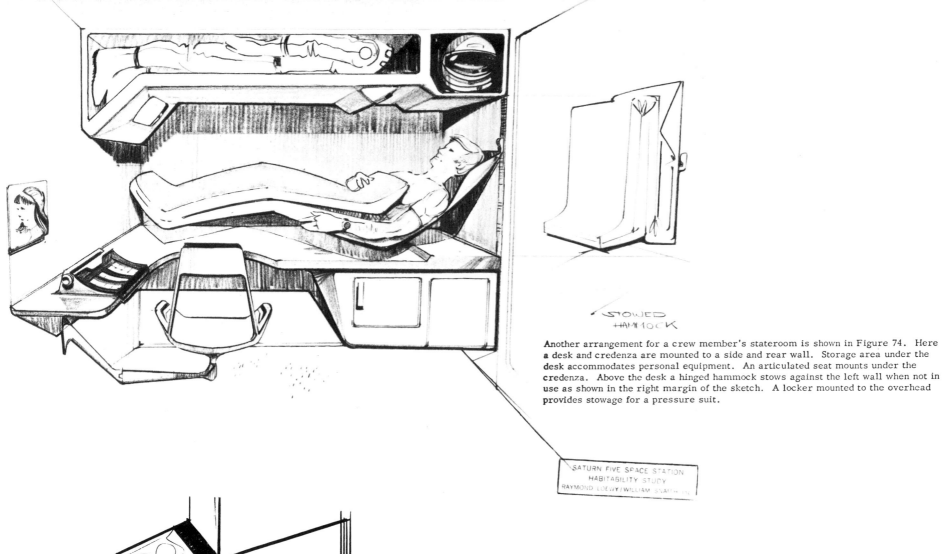

STOWED HAMMOCK

Another arrangement for a crew member's stateroom is shown in Figure 74. Here a desk and credenza are mounted to a side and rear wall. Storage area under the desk accommodates personal equipment. An articulated seat mounts under the credenza. Above the desk a hinged hammock stows against the left wall when not in use as shown in the right margin of the sketch. A locker mounted to the overhead provides stowage for a pressure suit.

SATURN FIVE SPACE STATION
HABITABILITY STUDY
RAYMOND LOEWY/WILLIAM SNAITH Inc.

SEAT SWINGS OUT FROM CAVITY.
WRITING TABLE DROPS DOWN.
SUIT RACKS OUT FROM STORAGE.

Figure 81. Consideration should be given to discarding the concept of installing a separate and distinct sleep station, be it horizontal or vertical. While, in our opinion, there is a valid psychological rationale for providing a separate area for sleep closely approximating conditions on Earth, from a purely functional viewpoint an adjustable seat could be made to offer equal comfort. In this sketch we show the seat being used during duty hours in conjunction with a writing surface that folds up from the storage wall. For sleeping, the headrest, back and seat panels could be adjusted to conform to what the crew member determined to be his most comfortable body attitude in zero G.

DOORS IN OPEN POSITION

SATURN FIVE SPACE STATION
HABITABILITY STUDY
RAYMOND LOEWY/WILLIAM SNAITH Inc.

Another combination work chair/sleep station is shown in Figure 82. In addition to a normal abdominal seat belt, shoulder, forearm and foot restraints are provided. During non-sleep periods, this chair would serve a writing surface that hinges down from the storage wall.

3.4C Waste Management Compartment

In Figure 89 we illustrate a conceptual arrangement for the WMS. The compartment is subdivided so as to offer two private fecal collectors, the space between used as a shower or body wash stall. In the open area, two personal hygiene cabinets are fixed, fitted with body restraints.

RESTRAINT

PH
UNIT

FECAL
STORAGE

WIPES

ASTROVAC

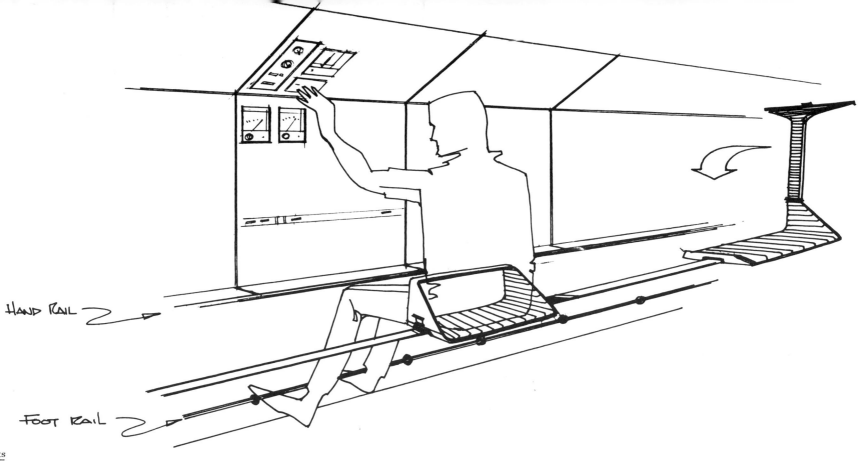

HAND RAIL

FOOT RAIL

3.4D <u>Restraints</u>

Figure 91. To provide a means of mobile restraint in the experimental and control areas where a crew man might be required to monitor a bank of consoles, we are suggesting this sliding seat. Fitted to a stationary beam, the seat could be slid from one station to another. Restraint is by means of a hinged, locking bar shown open at the right. Feet are restrained under another continuous beam. This toe restraint bar is fitted with protuberances at intervals to allow movement from station to station by providing purchase points for the feet.

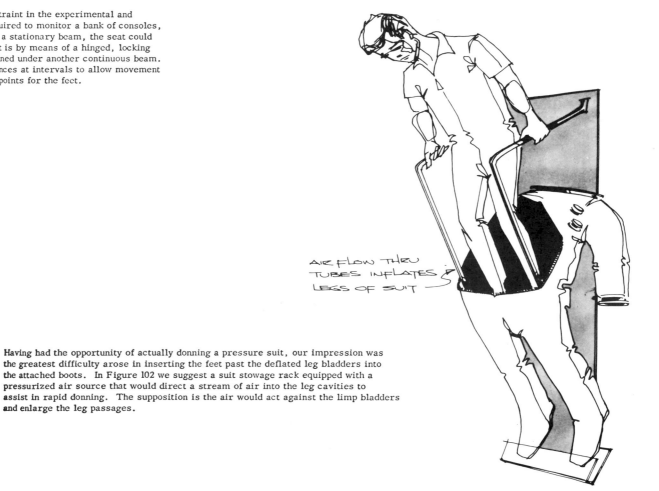

AIR FLOW THRU TUBES INFLATES LEGS OF SUIT

Having had the opportunity of actually donning a pressure suit, our impression was the greatest difficulty arose in inserting the feet past the deflated leg bladders into the attached boots. In Figure 102 we suggest a suit stowage rack equipped with a pressurized air source that would direct a stream of air into the leg cavities to assist in rapid donning. The supposition is the air would act against the limp bladders and enlarge the leg passages.

215

Exploded view of *Skylab* in space after components separation and reassembly for operational activity.

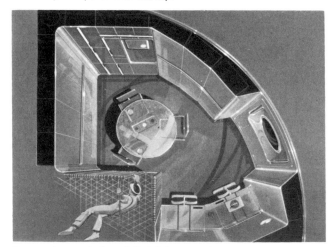

Perspective rendering of *Skylab* in operation.

Inside *Skylab*.

To Mr. and Mrs. Raymond Loewy —
— with admiration and appreciation for enriching the life style
of our world through your imaginative designs. The ASTP Crews.
(APOLLO-SOYUZ TEST PROJECT

Tom Stafford
Vance Brand
D. K. Slayton

NASA
HOUSTON TEX
APR 3 — 1975

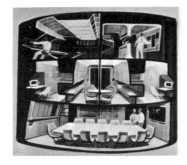

ivity, nuclear-powered space station.

In-flight combination of white, fireproof cloth with vertical Velcro strips. The overalls were conceived in close collaboration with NASA's research-and-development space-medicine department. Small items such as pencils, flashlights, screwdrivers, etc., are designed to adhere to Velcro. Sandals are designed for anti-levitation; the separate toes enable crew members to insert feet in the grillwork floor.

CKY STRIKE

March, 1940, George Washington Hill
ked into my office unannounced and said,
u Raymond Loewy?'' I said, ''Yes, I'm
mond Loewy.'' He then took off his jacket,
t his fishing hat on, sat down, and threw a
ack of Lucky Strikes on my desk.
I'm from American Tobacco.'' (He was the
sident.) ''Someone told me that you could
gn a better pack, and I don't believe it.''
en why are you here?'' I asked. He looked at
for a moment, grinned, and we were friends.
hout further ceremony, he pulled an attractive
arette case out of his pocket. ''Cartier,'' he
. ''Only the French can make this. And look
hese suspenders! Cartier, too.''
So are these,'' I said, showing my own, which
tier had made for me.
Well,'' he said, ''what about that package?
you really believe you can improve it?'' ''I bet
uld,'' I answered.
Ve did; we bet fifty thousand dollars. He left,
in April the new Lucky pack was adopted,
resulting sustained large sales increases,
ating at the same time a new look for
arette packaging. On the old green pack, the
ky Strike red circle (the target) appeared on
one side. Knowing they sold in the millions,
cided to display the target on *both* sides so
name Lucky Strike would be seen twice as
n. I replaced the green with a shiny white
the pack looked more luminous; it was also
aper to print and the smell which the green
had given off was gone.

Polyhedron concept.

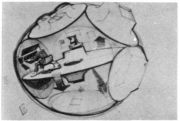

Concentric spherical concept.

Waste management research-and-development. Motion analysis of the defecation processes in three-dimensional mock-ups. Samples of waste matter from each astronaut can be collected in containers marked with crew member's name and time. Sealed container is then dehydrated, frozen, and stored, ready for analysis after debriefing.

Waste-management unit mock-up.

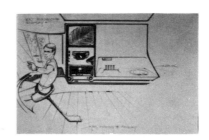

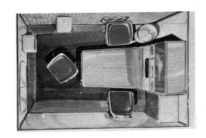

Interior studies.

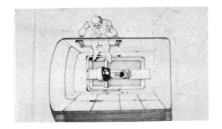

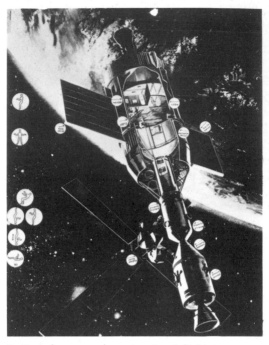

Position of crew members at various inflight stages.

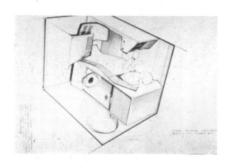

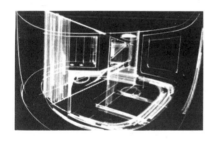

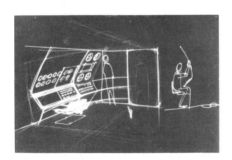

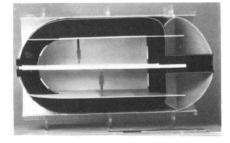

Interior space utilization and equipment studies.

Sketches for future spacecraft.

Folding seat with body and feet restraints.

I painted this moonscape from astronauts' impressions.

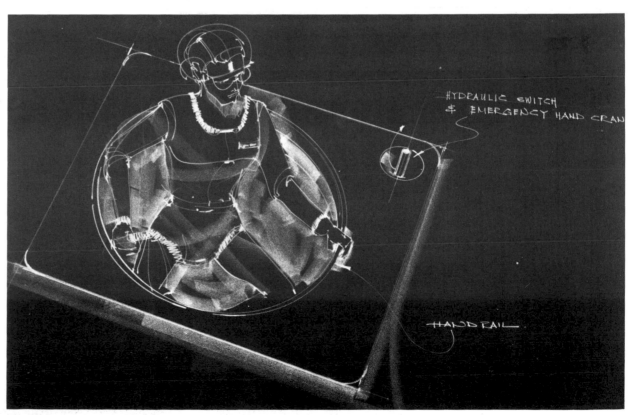

Inflight motion study.

221

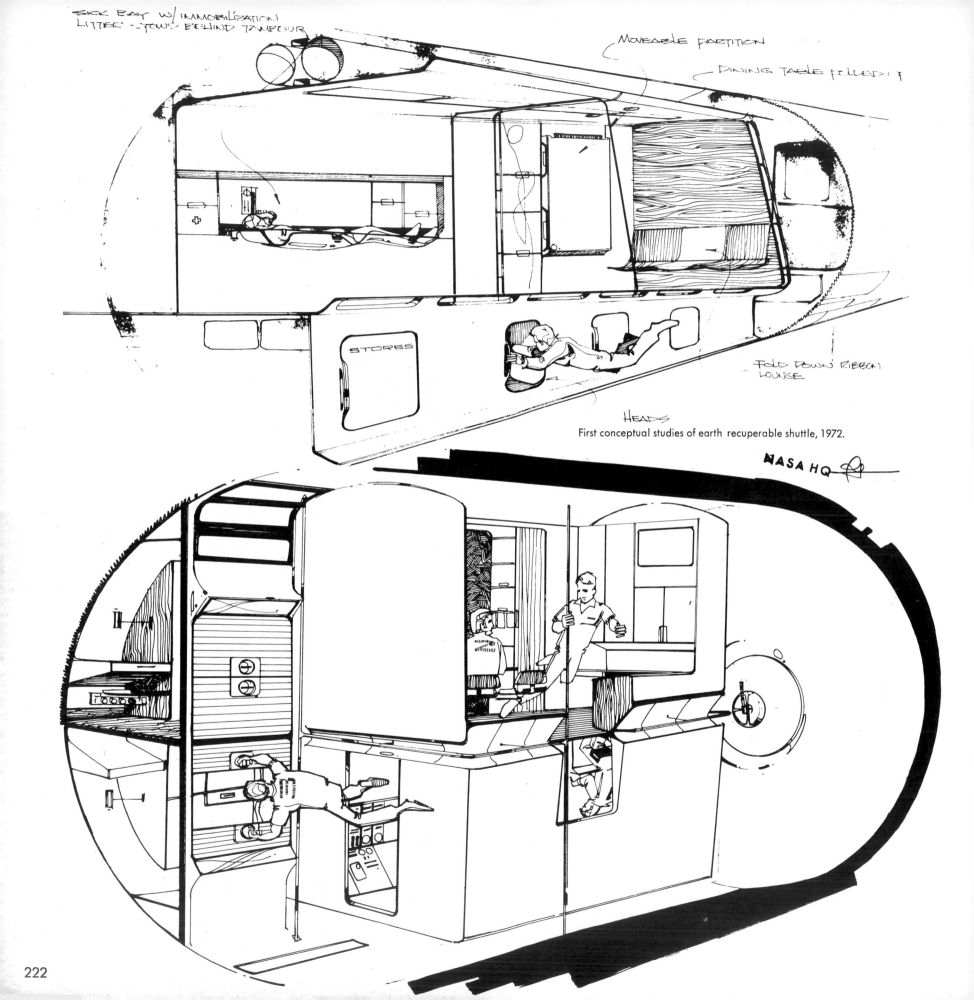

SICK BAY W/ IMMOBILIZATION
LITTER — TOWS BEHIND TAMBOUR

MOVEABLE PARTITION

DINING TABLE FOLDED UP

STORES

FOLD DOWN RIBBON
LOUNGE

HEADS

First conceptual studies of earth recuperable shuttle, 1972.

NASA HQ

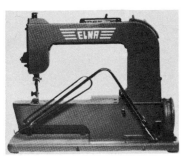

This Elna, 1964, is an exceptional quality "free-arm" portable sewing machine built in Switzerland by Tavaro. High precision, compactness, and light weight are the machine's features. It is shown here closed and open. The Elna is part of the international collection of contemporary mass-manufactured objects in the Museum of Modern Art in New York City. It was designed in Paris by our French company working closely with Tavaro's engineers.

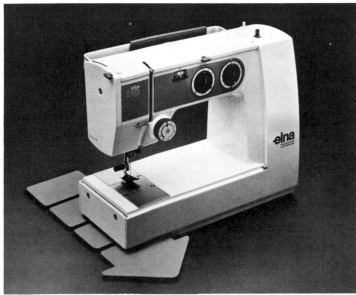

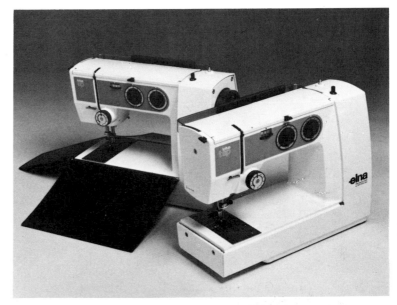

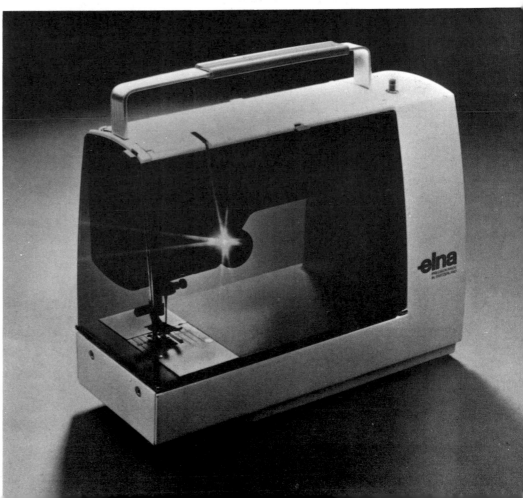

Pitney-Bowes postage metering machine, 1963.

The assignment was to design a jukebox that was both aesthetically appealing and highly marketable. We were told that it should be designed so that "smashed" customers could beat on it without its getting smashed, too.

Work gloves designed by C.E.I. that stressed good looks with no sacrifice of durability.

226

More than two decades ago, we conceived, planned, and designed the Waldorf Astoria's Bull and Bear restaurant in New York, which remains as we designed it, a popular Manhattan haunt for business people.

The Paris Hilton uniforms designed by C.E.I. The doorman's satchel contains coins for taxi riders who need change.

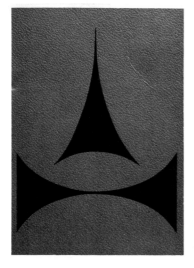

In 1966, C.E.I. planned the logo, layout, lighting, and décor of bedrooms and public areas for both the Orly and Paris Hiltons, as well as the furniture, lighting fixtures, flatware, glasses, dishes, menus, towels, and uniforms. I thought Parisians would enjoy the format of an American restaurant; Hilton called it The Western. We found much of the authtentic western ornamentation (cigar-store Indian, six-shooters, saddles, spurs, etc.) in Cabazon, California, a small desert town. The mobile was made from rusted handcuffs and spurs, stirrups, old guns, and sheriff's badges. Other touches were a long bar, shirt-sleeved, sideburned bartender, a large, bad nude of a fat lady reclining on a purple couch . . . all the clichés. And more: even a rinky-dink piano player, steaks branded with customer's initials, etc. Corny, yes; but what the client, from a business point of view, correctly wanted.

The Western's trademark, flatware, glassware, etc.

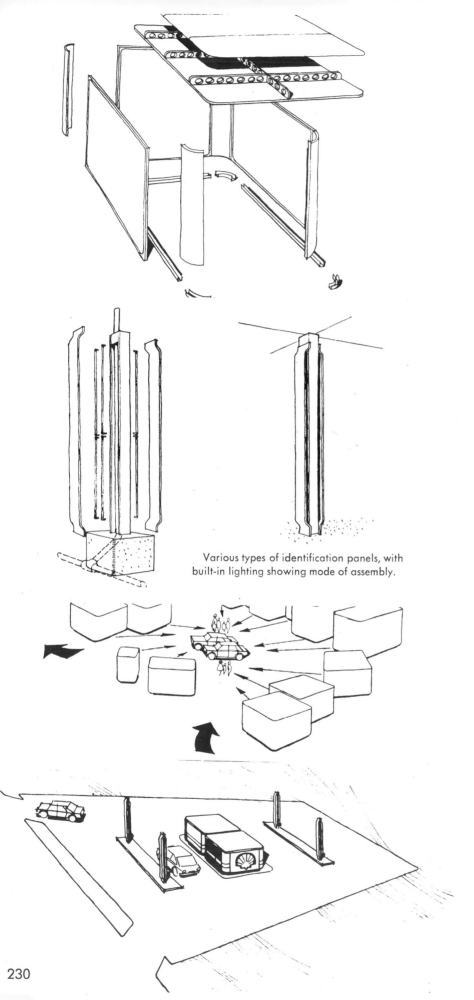

Various types of identification panels, with built-in lighting showing mode of assembly.

A close-up of the prefab modular system built slightly off the ground for better thermal insulation and easy maintenance.

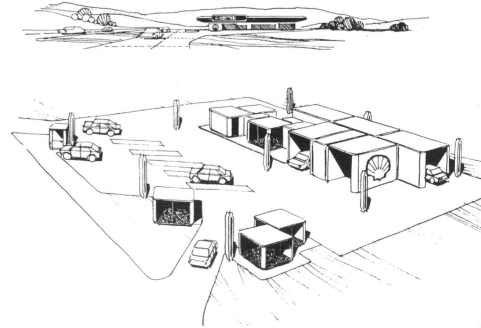

This MAYA prototype, erected on a Turin main thoroughfare, was such a success that, despite the station's large size, the station manager had to assign a man to organize lines. During the first years of operation, a station located two blocks away and operated by another major oil company had to suspend operations due to Shell's overwhelming competition. In the foreground, the service bays.

These illustrations show the versatility and adaptability of the MAYA (Most Advanced, Yet Acceptable) station prototype developed by C.E.I. for Shell International.

The radically different MAYA concept led C.E.I. into architectural engineering. The structure of the station is based upon a prefabricated module system. We worked directly with the engineering and marketing divisions of Shell International in their London headquarters. C.E.I. supervised the actual construction of every station.

Studies for an alternate MAYA prototype also based upon a prefab modular system. This version is called the "Pitch Roof Philosophy."

Our self-serve concept was an innovation that created a stir at the time among "the majors." C.E.I. designed it so that motorists could now refill a car's tank without the hose touching the ground. This eliminated soilage, the main consumer objection to self-serve. Pump traffic was designed for maximum order; the lighting was sunny and warm.

These trademarks were designed by C.E.I. The New Man trademark is reversible, equally readable when displayed up or down. Try it.

A trademark for Vachette, a French manufacturer of locks and bolts.

C.E.I. trademarks for canned goods, Vitos lingerie, and Omnium Plastic.

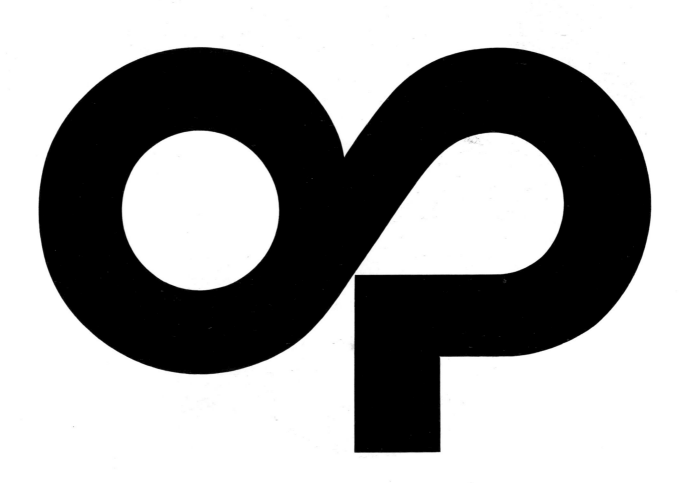

A trademark for a French manufacturer of chemical products, originally founded by Alfred Nobel, the inventor of dynamite. It now produces a diverse line of plastic products.

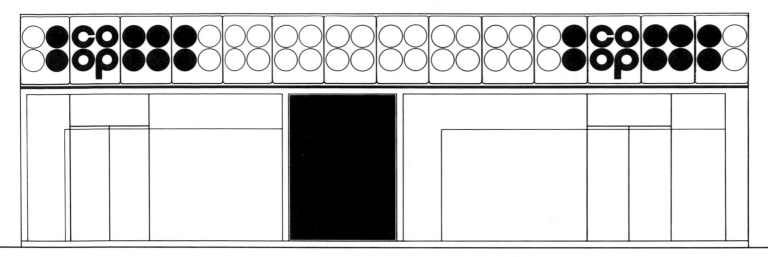

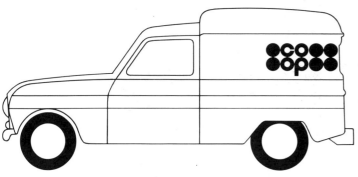

COOP is a large chain of French food shops and supermarkets. C.E.I. designed a prototype store that was adopted for 6,800 outlets. The trademark was applied to all kinds of vehicles, packaging, uniforms, letterheads, and advertisements.

Trademark for a leading manufacturer of chlorine-based disinfectants.

De Dietrich range and
heating equipment, 1974-75.

Package design for *petit beurre*,
the well-known biscuit
of Lefèvre-Utile, 1959.

Sprinter, the Dutch Railroads' new quick-accelerating,
high-speed train. Prototype was developed in scale-model
form by C.E.I. in 1974.

Prototype scale models for the Bertin aerotrain, 1970. These
fast trains glide over a cushion of compressed air. The ferryboat *Seaspeed* operates on the same principle.

Award-winning sea-going dredge, designed and built in America, 1974.

C.E.I. prototype for the Tridim, Bertin's aerotrain, 1969.

This silent switch, three inches wide, is easy to locate in total darkness; several can be clustered together and installed flush. When made of slightly flourescent plastic, they are even easier to locate at night. The success of the Arnould switch has led to C.E.I.'s developing a broad range of consumer products.

ASTER-BOUTILLON

A new type of electronic fuel-pump gauge. Its smooth surface makes it easier to clean, its pure lines subject it to less design obsolescence. It was designed by C.E.I. for one of Schlumberger's affiliated companies.

C.E.I. collaborated with Air France engineering for the interior design of the Concorde in 1973. We designed the reception area, the cabin's lighting, décor, and seats, as well as the flatware, china, glassware, and meal-service accessories. Unfortunately, passengers "liberated" several hundred sets of flatware as souvenirs during the first months of the plane's operation! Among our most challenging problems regarding interiors was to convey a visual appearance of width to the narrow fuselage. We decided to place a wide, jet-black, longitudinal band down the center of the ceiling, running the entire length of the aircraft. The band produced an optical illusion that suggested a wider fuselage.

A small part of this black center band is visible.

Another problem was the design of the headrest; because of fast takeoffs, passengers' necks must be supported by soft headrests. The seats are covered with fabrics of different colors interspersed without established rhythm or repetition. This gives the fuselage a visual diversity that a single color scheme would not have achieved. We sought to provide Concorde passengers with a cheerful interior amplified by warm lighting.

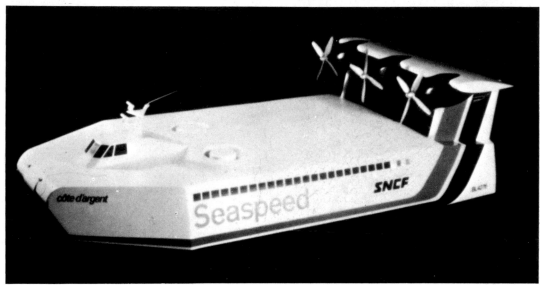

C.E.I.-built scale models. An actual photograph of *Seaspeed* in service. This type of craft can dock off beaches, a great advantage in countries lacking harbor facilities.

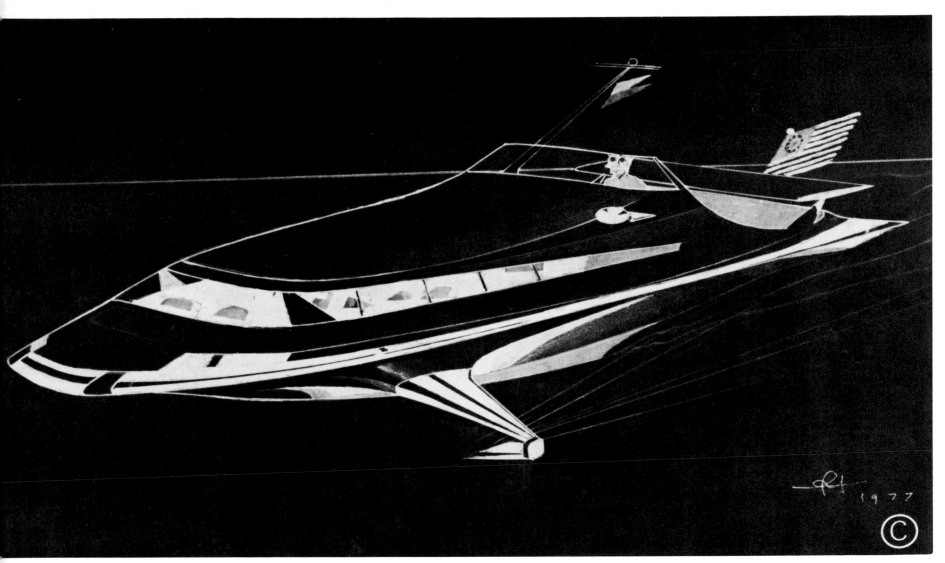

A smaller, faster hydrofoil, based upon design research done when we designed a 250-passenger commercial hydrofoil for the Russians. The angle of incidence of the horizontal plane at the stern can be modified to adjust the gliding angle, a functional design improvement.

Car of the future?

Selected List of Loewy Clients,
Products and Product Lines, and Projects

(In addition to this list being selective in itself, excluded for reasons of length are agencies
and divisions of government, public institutions, etc., from all over the world.)

Abraham & Straus
ACF Industries
Admiral
Aerotrain
Aireon
Air France
Albert Heijn
Albert Parvin
Alfa Laval
Allied Breweries
Allied Chemical
Amana Refrigeration
Ambassador Scotch Whisky
American Banner Lines
American Brake Shoe
American Can
American Car & Foundry
American Cyanamid
American Export Isbrandtsen
American Express
American Hard Rubber
American Hardware
American Locker
American Machine & Foundry
American News
American Oil
American President Lines
American Rocket Society
American Safety Razor
American Secretariat
American Telephone & Telegraph
American Tobacco
Andrew Jergins
Angelique
Ansco
Ansul Chemical
A.O.I.P.
Aqua Tec
Arcadia Fisheries
Arketex Ceramic
Armour
Arnould
Artcraft Venetian Blind
Arvin Industries
Asahi Breweries
Associated Dry Goods
Atlas Chemical Industries
Audemar
Auer
Austin Motors
Automatic Canteen
Automatic Radio
Avanti
Avco
Avions Marcel Dassault
Azur Plastique

Baignol & Farjon
Baldwin Locomotives
Baranne
Baroclem
Baumgartner
Beckman Instruments
Beechams
Belin
Bell Aerospace
Bell Aircraft
Ber-Design
Berkel
Bernardaud
Bertin
Best & Co.
Best Foods
Bières Basse Yutz
Biltmore Hotel
Birmingham Sound
Biscottes Saint-Luc
Blanchaud
Bloomingdale's
Blue Bell
Boeing Airplane
Bonwit Teller
Borg-Erickson
Bowater Scott
Bremshey
Bristol-Myers
British Aircraft
British International Paper
British Motor
British Petroleum
Britvic
Broadway-Hale Stores
Broil-Quik
Brown-Forman Distillers
Brown & Polson
Brunswick-Balke-Callendar
Brush Instruments
Brush-O-Dent
BSR
Buitoni
Bulova Watch
Butterick Publications

Cadbury Chocolates
Caem-Mercedes
Caloric
Campari
Campbell's Soups
Canada Dry
Canadair
Canadian National Railway
Canadian Pacific Railway
Carillo Coffeemakers
Carling Breweries
Carrier Air Conditioning
Carteret
Carter Products
Castel-Joyeux
Ca-Va-Seul
Celanese
Celotex
Central Rubber & Steel
CGE-FAEM
Chaffoteaux et Maury
Chandler-Evans
Charmilles
Chicorée Leroux
Chrysler
Chubb
Cigarettes Laurens
C.I.M.
City Stores
Claude, Paz & Visseaux
Clement Gaget
Cleo
Clopay
Coca-Cola
Cockshutt Farm Equipment
Cognac Martell
Cointreau
Colgate-Palmolive
Colonial Radio
Computer Terminal
Concorde
Congespirin
Conord
Consolidated Aircraft
Contigea-Buhler
Continental Baking
COOP
Cooper Alloy
Cormac Photocopy
Cornell-Dubilier Electronics
Corona
Cotelle & Foucher
Country Togs
Cramer Controls
Crompton & Knowles
Cuenod

Daichi Busson Kaisha
Dayton Rubber
Dayton's
De Bijenkorf
De Dietrich
Delaware & Hudson Railroad
Délico
Delta Electric
Despar
Diamond
Dictaphone
Dictograph Products
Diebold
Diehl
Documat
Douglas Aircraft
Dow Chemical
Drackett
Dröste
Drummond
DuPont, E.I., de Nemours
Dynatra

Eagle Food Centers
Eagle Pencil
Echo de la Mode
Ed. Schuster & Co.
Ekco Products
Electric Boat
Electro Mechanical Research
Elgorriaga
Emerson Radio
Emporium
E.M.R.
Enger-Kress
Englebert-Uniroyal
Equitable Life Insurance
E.R. Squibb & Sons
Essex House Hotel
Esterline-Angus Instrument
Euromarche
Everitube
Eversharp
Exposition Internationale de Bruxelles
Exxon

Fabergé
Fairbanks, Morse
Fairbanks Whitney
Fairchild Engine & Airplane
Fairchild Recording Equip.
Famous-Barr
Fanta Beverage
Fasan Durasharp
Fedders
Federal Enameling & Stamping
Federal Pacific Electric
Federated Dept. Stores
Feudor
Fichet-Bauche
Filene's
Firestone Tire & Rubber
First National City Bank
Fobrux
Foley Brothers
Ford Motor
Formica
Fostoria Glass
Franklin Aluminium
Fred Harvey
Frigidaire
Frimotor-Westinghouse
F. & R. Lazarus
Fruidor
F. Schumacher
Fujiya Confectionery

Galleries Lafayette
Garnier
Gayfer's
General Aniline & Film
General Concentrates
General Electric
General Fireproofing
General Foods
General Motors
General Time
Gestetner
Gillette Safety Razor
Gimbel's
Glenlivet Distillers
Glenn Martin Aircraft
Godin
Gold Seal
Goodyear Tire & Rubber
Greyhound
Grey-Poupon
Grumman Aircraft Engineer
GTO Records
Guigoz
Guinness

Hahne & Co.
Halle Bros.
Hallmark Cards
Hanes Hosiery
Hanovia Chemical
Haus Neuerburg
Heinz
Hénaff et Fils
Henry Morgan & Co.
Hess Chemical
Higbee's
Higgins Boat
Hills Bros. Coffee
Hilton Hotels
Hoover
Horn Convecteurs
Horten
Hotpoint
Howard Hughes Aircraft
Huber
Hudnut-DuBarry
Hupmobile

I.B.M.
Ideal Standard
Ideax
IDV
Imperial Desk
Independent Telephone
Indian Head Hosiery
Intercontinental Hotels
International Harvester
International Paper
Interstate Motor Freight Sys
ITM/Capital Tea

Jacobs Kaffee
Japanese Tobacco Monopoly
J.A. Ragland's
J.B. Ivey & Co.
J.B. White
J.C. Penney
J. Horne & Co.
J.J. Newberry
J.L. Hudson
Job
John Bressmer
John Lewis
Johns-Manville
Johnson & Johnson
John Wanamaker
Joseph Magnin
Joy Mfg.
J.W. Robinson

Kaiser Aluminium & Chemical
Karstadt
Kelvinator Fooderama
Kennedy Foundation
Kimberly-Clark
Klein's
K.L.M. Airlines
Knorr
Koehler Mfg.
Kroger
Kuhn, Loeb & Co.

Laboratoires Roussel-UCLAF
Lancia
Landers, Frary & Clark
Landmark Farm Cooperatives
L. Bamberger
Lebaudy Sommier
Le Creuset
Le Dauphin
Lefèvre-Utile
Lesieur-Cotelle
Lever Bros.
Levis
Lewisburg Furniture
Liebig
Liebmann Breweries
Lightolier
Lily-Tulip Cup
Limoges-Castel
Link Aviation
Lockheed Aircraft
Long Island Railroad
Looza
Lord & Taylor
Lorraine Bauer
L.S. Ayres
Lucent
Lucky Stores
Lummus
Lustucru

Maille
Mandel Bros.
Manning, Maxwell & Moore
Marina America
Mars Confectionery Works
Martougin
Marumiya
Marwell Construction
Masonware
Massey-Ferguson
Matson Lines
Mattel
May Dept. Stores
McCray Refrigerator
Meier & Frank
Menier
Mercantile Stores
Merlite Industries
Mesberg
Metric Hosiery
Metropolitan Builders
Michael Reese Hospital
M.I.C. Transpallette
Miko
Miller & Rhoades
Minneapolis-Honeywell Regulator
Missouri-Pacific Railroad
Mitsui
Mobil Oil
Monoprix
Monorail Rapid Transit System
Monsanto
Monsieur Henri Wines
Montreal Museum of Fine Arts
Moore-McCormick Lines
Morley Furniture
Moskvitch Automobiles
Mosler Safe
Motta

Nabisco
NASA
Nashua
National Airlines
National Boat
National Brush
National Dairy Products
National Distillers & Chemical
NBC
Nesco
Nestlé
Neue Warenhaus AG
New Man
Newsweek
New York Telephone
Niemeijer
Nobel-Bozel
Norcross
Nord Aviation
Norfolk & Western Railway
Normacem
North American Van Lines
Northeast Airlines
Northern Pacific Railway
Northrop
Northwest Aeronautical
Nova
Noyama
Nutone
Nutrilite Products

O'Cedar
Oertel Brewing
Ogilvie Flour Mills
Ohrbach's
O'Keefe's Brewing
Omar
Omega Watch
Orchard King
Oreal
Origny Sugar
Owens-Corning
Owens Staple-Tied Brush

Panama Lines
Pantasote
Parein
Parfums Weil
Paris-Rhone
Paymaster
Peace Corps
Pearson Pharmacal
Peck & Peck
Pennsylvania Railroad
Pepsodent
Perret Jaunet
Perrochet
Pharmacraft Laboratories
Philadelphia Sunday Bulletin
Philip Morris Tobacco
Phillips Petroleum
Picker X-Ray
Pied-Selle
Pioneer Rubber
Pitney-Bowes
Plastic Omnium
Plastimonde
Plymouth Shops
Poulain
Priba
Printemps
Procter & Gamble
Proctor-Silex
Purity Stores

Quaker Oats

Rank Hovis McDougall
Reddi-Whip
Renault
Republic Aviation
Restaurant Associates
Revlon
Rheingold Breweries
R.H. Macy
Rhone-Progil
Rich's
Rival Mfg.
R.J. Reynolds Tobacco
Roberk
Rod Pickard Yachts
Roehr Products
Rogers Brothers
Rogers Peet
Romanet
Roneo
Rootes Motors
Rosenthal-Block China
Rotet-Pano
Royal McBee
Rudd-Melikian

Sable
Saks Fifth Avenue
Sanger-Harris
Sarma
Sartel
Savannah Sugar Refining
Schenley Industries
Schick
Schieffelin & Co.
Schlage Lock
Schneider
Scott-Atwater
Seagram Distillers
Sealtest
Sears, Roebuck
Seita
Seneca Textile
Servo
Seth Thomas Clocks
Shell International
Sikorsky Aircraft
Simca-Someca
Simmons
Singer Sewing Machine
S.N. Pétroles d'Aquitaine
Southern Pacific
Spar International
S. & S. Corrugated Paper
Standard Brands
Standard Cap & Seal
Standard Motor
Standard Oil
Standard Packaging
State Mutual Life
Steinberg's
Stewart Dry Goods
Stix-Baer & Fuller
Stouffer
Studebaker-Packard
Studio Opéra
Suchard Chocolate
Sud-Aviation
Sunbeam
Sun Oil
Super Market Institute
Sure Save
Swivelier
Synergie

Tata Air Lines
Tavaro/Elna
Taylor Instrument
Tekni-Craft
Teletype
Terraillon
Testut
Texaco
Textron
Thompson Products
Thor Power Tool
Threshers
Tintair
Titmus Optical
Tom Thumb Super Markets
Torsion Balance
Total Oil
Tracerlab
Transair
TWA

Uhu Werke Fismar
Ulis Supermarche
Uncle Ben
Unicliffe
Unilever
Union Carbide
Union de Brasseries
Union News
Union Switch & Signal
Unipol
United Aircraft
United Air Lines
United Music
United Parcel Service
U.S. Steel

Vandemoortele
Vander Elst
Vanguard
Vendo
Verzinkerei Zug
Viandox
Viking Air Conditioning
Virginia Ferry
Vitos
Viva
Volvic
Vulcan Radiator
Vultee Aircraft
Vynckier Frères

Wagons-Lits
Waldorf-Astoria
Wallace Silversmiths
Waring Products
Weber & Heilbroner
Weber Showcase & Fixture
Welle
Wellington Sears
Western Engineering
Westinghouse Electric
West Virginia Pulp & Paper
Weyerhauser
Wheeling Corrugated
Whitbread
White-Rogers
Whitney Blake
Wood Conversion
Woodward & Lothrop
W. T. Grant
Wyeth Laboratories

Zippo

249

SELECTIVE INDEX OF ILLUSTRATIONS

RHM ◖◗◖◗◖◗
RANKS HOVIS McDOUGALL LIMITED

 AIR **BP**

 NEW MAN

 **viandox**

CABY

ĕcho
de la mode

 **PROTEL**

 **De Dietrich**

 **PÉPRO**

DIAMOND TRUST

 **DEHO SYSTEMS**

VIRLUX

Sauveterre

ROUSSEL UCLAF

NOORD BRABAND NB

co op

AMERICAN ★ CHAM... OF COMMERCE IN FR...

OP
PLASTIC OMNIUM

 E L

elna

corona

V

 **QUAKER**

SPAR